Lecture Notes in Physics

344

M. Rascle D. Serre
M. Slemrod (Eds.)

PDEs and Continuum Models of Phase Transitions

Proceedings of an NSF-CNRS Joint Seminar
Held in Nice, France, January 18–22, 1988

Springer-Verlag
Berlin Heidelberg New York London Paris Tokyo Hong Kong

Editors

M. Rascle
Département de Mathématiques, Université de Nice
Parc Valrose, F-06034 Nice Cédex, France

D. Serre
Ecole Normale Supérieure de Lyon
46, allée d'Italie, F-69364 Lyon Cédex 07, France

M. Slemrod
Mathematics Department, University of Wisconsin
Walnut Street, Madison, WI 53706, USA

ISBN 0-387-51617-4 Springer-Verlag New York Berlin Heidelberg
ISBN 3-540-51617-4 Springer-Verlag Berlin Heidelberg New York

Printed in Germany

Printing: Druckhaus Beltz, Hemsbach/Bergstr.
Bookbinding: J. Schäffer GmbH & Co. KG., Grünstadt
2153/3140-543210 – Printed on acid-free paper

PREFACE

The N.S.F. - C.N.R.S. Joint Seminar : "Partial Differential Equations and Continuum Models of Phase Transitions" was held in Nice in January 1988, under the auspices of the International Society for the Interaction of Mechanics and Mathematics and within the framework of the 1987 - 1988 "Année spéciale sur les phénomènes non linéaires", supported in France by C.N.R.S. and the Ministère des Universités.

The study of phase transitions, the importance of which is crucial in many physical applications, is obviously one of the fundamental problems of Physics. During the last few years, spectacular progress has been made in constructing and understanding better continuum models and the associated systems of partial differential equations, in spite of the difficulty of problems such as the existence of solutions to "wildly" non-convex problems or the Hadamard instability in mixed-type systems. Therefore the goal of this Seminar was to bring together scientists from different fields (applied mathematics, theoretical or computational mechanics, solid-state physics, etc). Through a few examples (micro-structure of crystals, defects in liquid crystals, liquid-vapor interfaces, etc), the speakers have described several key points of continuum models (structure and evolution of interfaces, regularization by means of interfacial energy, (non-)equilibrium theories, etc) and some of the most typical mathematical problems and results: study of oscillations (Young's measures), singular solutions, admissibility of shock waves, role of higher-order perturbations, etc.

This Seminar had a very pleasant atmosphere, and we think it was highly successful. We really enjoyed the quality of the talks and the contributions in this volume as well as the opportunity of developing scientific contacts and cooperation. We are particularly grateful to John BALL, James GLIMM, Morton GURTIN, Maurice KLEMAN and Gérard MAUGIN for their help in organizing this workshop.

Finally we are particularly grateful to the following organizations which have supported this meeting:

C.N.R.S., N.S.F., S.M.A.I., G.A.M.N.I., D.R.E.T., the European Research Office of the U.S. Army, le Conseil Général des Alpes Maritimes, l'Université de Nice, and we wish to thank for their assistance:

Marie-Christine BERMOND, Annie BOREL, Sonia CHAKOFF and Janine LACHKAR.

Nice, September 1989

Michel RASCLE
Denis SERRE
Marshall SLEMROD

PREFACE

Le Séminaire conjoint C.N.R.S.-N.S.F: "Equations aux dérivées partielles et modèles continus de transitions de phases" s'est tenu à Nice du 18 au 22 janvier 1988, dans le cadre de l'Année spéciale 1987 - 1988 sur les phénomènes non linéaires, organisée en France sous le haut patronage du C.N.R.S. et du Ministère des Universités. Ce Séminaire était organisé sous les auspices de l'International Society for the Interactions of Mechanics and Mathematics.

L'étude des transitions de phases est évidemment l'un des problèmes fondamentaux de la physique, d'importance cruciale dans de très nombreuses applications. Au cours des dernières années, des progrès spectaculaires ont été réalisés dans l'élaboration et la compréhension des modèles continus et des systèmes d'équations aux dérivées partielles associés, qui posent des problèmes difficiles; par exemple l'existence de solutions à des problèmes variationnels "violemment" non convexes ou l'instabilité d'Hadamard pour des systèmes de type mixte. L'objectif de ce Séminaire était donc de permettre à des spécialistes d'horizons différents (mathématiques appliquées, mécanique théorique ou numérique, physique du solide...) de se rencontrer et de faire le point. A travers un certain nombre d'exemples (micro-structures des cristaux, défauts dans les cristaux liquides, interfaces liquide-vapeur etc...), les conférenciers ont traité plusieurs points-clé de la modélisation (structure et évolution des interfaces, régularisation via l'énergie interfaciale, théories thermodynamiques (hors)équilibre etc...) et présenté quelques-uns des problèmes et des résultats mathématiques récents les plus typiques: étude des oscillations (mesures d'Young), solutions singulières, admissibilité des ondes de choc, rôle des perturbations d'ordre supérieur etc...

Ce Séminaire s'est déroulé dans une ambiance très agréable et nous pensons qu'il a été très fructueux, tant par la qualité des exposés et des contributions de ce volume que par les contacts qu'il a permis de nouer et de développer. Nous tenons à remercier particulièrement John BALL, James GLIMM, Morton GURTIN, Maurice KLEMAN et Gérard MAUGIN pour leur participation à l'organisation de cette rencontre.

Enfin, nous remercions particulièrement pour leur soutien:

C.N.R.S., N.S.F., S.M.A.I., G.A.M.N.I., D.R.E.T., European Research Office of the U.S. Army, Conseil Général des Alpes Maritimes, Université de Nice,

et pour leur aide dans l'organisation matérielle de ce Séminaire:

Marie-Christine BERMOND, Annie BOREL, Sonia CHAKOFF et Janine LACHKAR.

LE PRESENT DOCUMENT A ETE ETABLI EN EXECUTION DU CONTRAT N° 87/1454 PASSE PAR LA DIRECTION DES RECHERCHES ET ETUDES TECHNIQUES, DIRECTION SCIENTIFIQUE, SECTION SOUTIEN A LA RECHERCHE (DELEGATION GENERALE POUR L'ARMEMENT).

CONTENTS

Part I

CONTINUUM MECHANICS

CLUSTERS OF SINGULARITIES IN LIQUID CRYSTALS

Yves BOULIGAND

Laboratoire d'Histophysique et de Cytophysique de l'Ecole Pratique des Hautes Etudes.
CNRS, 67 rue M.-Günsbourg, 94200 Ivry-sur-Seine (F.).

Fluid structures and textures

Liquid crystals are well known ordered media, offering beautiful pictures in polarizing microscopy (1-5). Molecules are elongated in general and align locally, whereas the medium is fluid. The birefringence is often high (6) and polarization colours appear, when such liquids, mounted between slide and coverslip, are observed between crossed nicols. The main *structures* of liquid crystals (7) are represented in Fig.1. In *smectic* liquid crystals, the parallel elongated molecules have their centres of gravity arranged on parallel equidistant planes, as can be seen from Fig.1a and the molecules often normal to layers are mobile within the smectic layers. In the *nematic* state, the parallel elongated molecules have their centres of gravity arranged at random, with however a constant density (Fig.1b). The *cholesteric* state is a variety of nematic phase with a helical arrangement of molecules leading to a periodicity (Fig. 1,c,d).

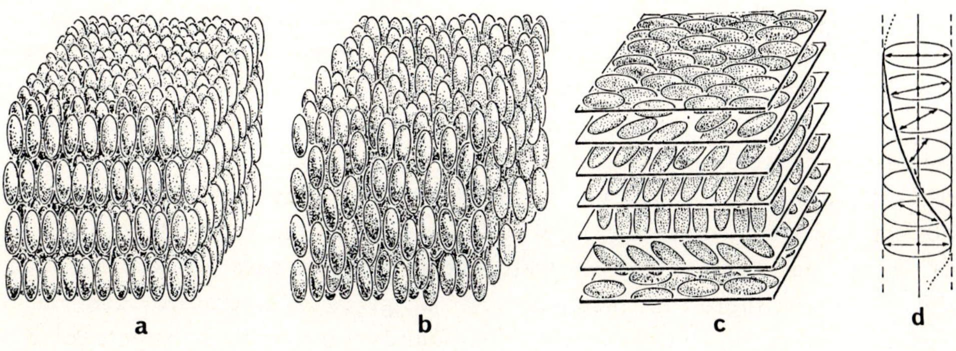

Fig.1. Molecular arrangement in the three main types of liquid crystals. a. In *smectic phases,* molecules are ordered in lamellae of equal thickness. b. In *nematic phases,* the centres of gravity of the parallel elongated molecules are randomly distributed. c. In *cholesteric phases,* the molecules rotate from plane to plane and are normal to an axis represented in d; the planes are drawn for convenience, the structure being continuous. d. Helix associated with the cholesteric arrangement.

The symmetries and mechanical properties of these media are compatible with continuous distortions and different types of curvature were recognized in liquid crystals (8,9). These distortions also lead to various obstructions, resulting in singularities, which are points, lines or walls (1,3,5). These 'defects' or singularities are the core of configurations of the surrounding medium, which cannot be deformed continuously into the liquid monocrystal, characterized by a perfect alignment. These configurations around singular points or lines are not trivial and belong to different homotopy classes (10-12). Singularities move easily, owing to the medium fluidity, and reach positions minimizing their elastic energies, when the liquid is at rest. The resulting distribution of defects is often regular (3). The medium distortions around singularities create patterns or *textures*, and examples are presented in Fig.2. In cholesteric liquids, series of similar patterns are repeated and form clusters, separated by zones where the liquid crystal shows a more uniform orientation, as in Fig.1a. The cholesteric defects of Fig.2b are less regularly distributed, but parallel equidistant lines are observed, the distance separating two lines corresponding to a rotation of 180°, as in Fig.1c. In smectics, clusters of focal conics are frequent (Fig.2c).

Regular textures are also observed in various hydrodynamic conditions and characteristic steady states were described, when liquid crystals are submitted to thermal and electric gradients (13,14) or to mechanical constraints such as alternating shears (15). The hydrodynamics of liquid crystals is mainly studied in aligned samples of mesophases, for different geometries used in these experiments and the mathematical theory is limited to these ideal conditions (16). Actually, liquid crystals are not aligned in general and the flow is extremely heterogenous, since defects may arrange into clusters.

Liquid . crystalline phases are observed in a lot of pure organic compounds and also in mixtures (6). Phase diagrams are complex and several different phases can be obtained at one point of the phase space but do not coexist in general, since one of them is the most stable, the other phases being metastable. This is well known in the case of simple systems showing supercooling, for instance. At a given temperature and at the atmospheric pressure, many pure compounds are either supercooled or crystallized. In liquid crystals, metastability is frequent and the observed phase depends on the history of the medium and more precisely on the path followed in the phase space. Texture diagrams are much more complex than structure diagrams, since they depend on the velocity of the phase transitions and on various factors, such as surface structure of the preparation glasses, external fields and hydrodynamic conditions.

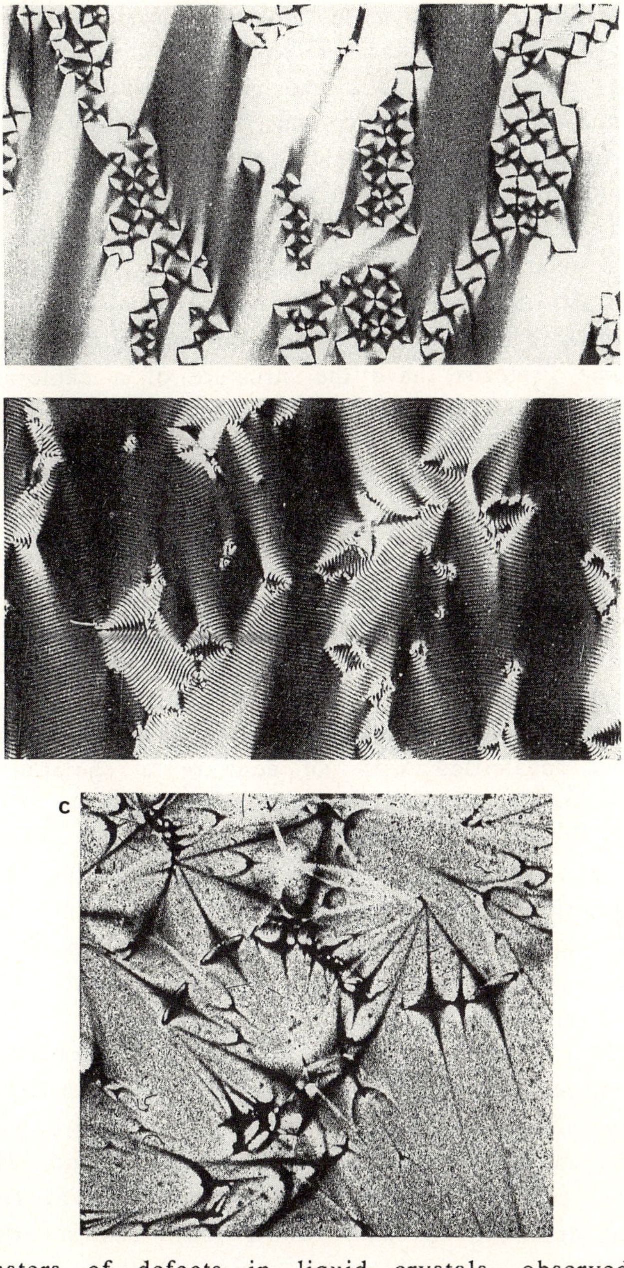

Fig.2. Clusters of defects in liquid crystals, observed in polarizing microscopy. a. Cholesteric liquid with uniform domains and polygonal fields. b. Cholesteric liquid with some of the patterns considered in Fig.3. c. Focal conics in a smectic liquid.

Liquid crystals are present in biological systems and are involved in morphogenesis (17). Cell membranes and their derivatives are well known examples. Most structures of cells and tissues present characters of ordered media with symmetries of liquid crystals, but are not fluid in general; they are stabilized analogues of liquid crystals. Such systems of aligned polymers or parallel elongated micelles are gelated by a set of cross-links. At the time of their elaboration, these materials are fluid and their textures are fixed by a gelation process. The analysis of liquid crystalline textures and their dynamics is therefore an essential step towards the study of biological morphogenesis. Chromosomes in primitive organisms, such as bacteria and some Protozoa, have the structure of a cholesteric liquid and it can be observed that purified DNA, sonicated into small fragments concentrated in water, can form cholesteric droplets, with the elongated shape of chromosomes (18).

Cinematographic studies of liquid crystalline textures

A movies was realized some years ago by Jean Painlevé on the parallels I had observed between biological materials and liquid crystals (19). The carapace of crabs is a composite material, made of organic fibrils and calcite crystals, which makes the system very hard. However, the carapace can be decalcified with an acid or a chelating agent, and the resulting organic matrix is supple and appropriate for the preparation of thin sections, after embedding in paraffin or resins. From their examination, it can be shown that these matrices have the structure of a stabilized cholesteric liquid crystal. This is also illustrated from parallel observations of sections in the matrix of crab carapaces and of nematic samples, twisted by the addition of a chiral compound and sandwiched between slide and coverslip. The cholesteric twist generates spiralized patterns in both materials, with similar optical properties in polarizing microscopy. However, thermal fluctuations are well evidenced in the liquid crystalline preparation, in phase contrast microscopy, whereas they are absent in matrix sections, since this material is not fluid, but stabilized. The main types of singularities and textures are presented in the film; they are well known and described in many works (20). Textures observed in liquid crystals are more or less stable and those of nematic liquid crystals are easily transformed by the presence of weak streams within the medium, since there are no strong coupling between the positions of defects. On the contrary, liquid crystals with discrete periodicities, such as smectics (layer thickness) and cholesterics (half helicoïdal pitch), have much more stable textures, and large areas remain almost unaltered, even in the presence of important flows. These domains contain defects associated into stable clusters. Dislocations, focal lines and disclinations either converge to singular points or locally superimpose, or present various geometric interrelations, such as those observed for instance between focal conics (3).

The domains characterized by such clusters of singularities move with few deformations and they are simply translated or rotated by the surrounding streams. Smectic or cholesteric layers align between these domains, what facilitate the displacement of large stable domains relative to other stable fields. The aligned layers lying between the unaltered domains are therefore submitted to various shears. Defects create a rigidity over certain areas of textures and clusters of geometrically associated defects resist the flow. However the apparently stable domains slowly transform through rearrangement, creation or anihilation of defects. When the flow is much more important, with strong viscous torques, the stable areas probably disappear.

Some of the associations of defects which stabilize textures are represented in Fig.3. Another method to investigate the stability of certain textures is to squash strongly and briefly the preparation. Certain textures are rapidly restored with their characteristic patterns, particularly those of Fig.3. Other textures redifferentiate after a longer delay, and this is the case of polygonal fields for instance, which are textures resulting from the association of focal lines into a lattice.

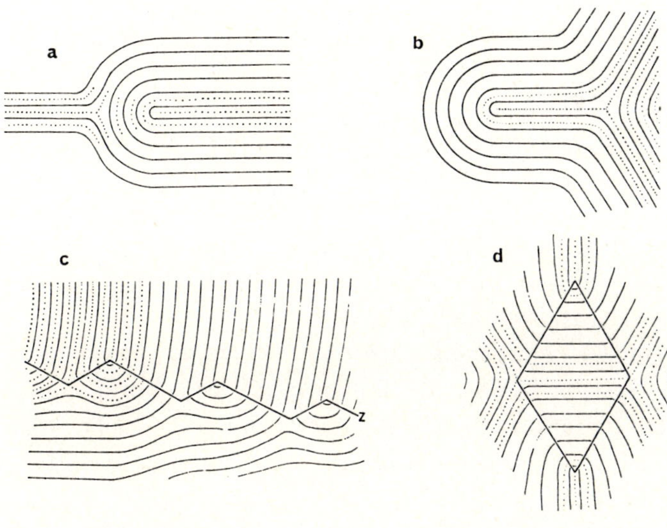

Fig.3. Main patterns observed in most stable textures of cholesteric liquids. a. Edge dislocation split into two disclinations. b. Elementary pinch. c. Zig-zag arrangement of disclinations. d. Quadrilateral of four disclinations.

The existence of stable clusters of defects, separated by sets of layers working as a lubrificant, give to liquid crystals, mainly those presenting discrete translational symmetries, an heterogenous texture, which is more or less granular. This is well known from experimentalists, who manipulate these liquids, in conditions we consider as 'natural', i.e., before experiment, when the liquid is far from being aligned, what is required at a given step in most works. Defects are often created by phase transitions, and their distribution depends on conditions realized at the transition and on further constraints. The hydrodynamical properties are sensitive to the presence of singularities, particularly in the vicinity of phase transitions and this makes difficult the elaboration of continuous models of symmetry breaking in such systems.

Literature

1. Lehmann O. *Die neue Welt der flüssigen Kristallen und deren Bedeutung für Physik, Chemie, Technik und Biologie*. Akad. Verlag Ges., Leipzig, 1vol., 388 p. (1911).
2. Friedel G., *Ann. Physique,* Paris, **9** (18), 273- (1922).
3. Bouligand Y. *J. Physique, Paris,* **33**: 715 (1972); **34**: 603, 1011 (1973); **35**: 215, 959 (1974).
4. Demus D. and Richter L., *Textures of Liquid Crystals,* VEB Deutscher Verlag für Grundstoffindustrie, Leipzig, (1978).
5. Kléman M. Points, Lines and Walls, 1 vol., 322 p., J. Wiley (1983).
6. Kelker H. and Hatz H., *Handbook of Liquid Crystals*. Verlag Chemie, Weinheim (1980).
7. de Gennes P.-G. *The Physics of Liquid Crystals,* Clarendon Pr. (1975).
8. Oseen C. W. *Trans. Faraday Soc.*, **29**: 883 (1933).
9. Frank F. C., *Disc. Faraday Soc.,* **25**: 19 (1958).
10. Toulouse G. and Kléman M., *J. Physique,* Paris, **37**: L-149 (1976).
11. Bouligand Y., Derrida B., Poenaru B., Pomeau Y. et Toulouse G., *J. Physique,* Paris, **39**: 863 (1978).
12. Poenaru V., in *Ill-condensed Matter*, Ecole de Physique Théorique des Houches, Session XXXI, R. Balian *et al.*, ed., North Holl., 263 (1979).
13. Guyon E. et Pieranski P. *C. R. Acad. Sci.,* B, **274:** 656-760 (1972). Pieranski P., Dubois-Violette E. et Guyon E., *Phys Rev Lett.,* **30**: 736-739 (1973).
14. Durand G., Gennes P.-G. de et Veyssié M., dans *Quelques Aspects de l'Etat Solide Organique,* J.-P. Suchet éd., Séminaires de Chimie de l'Etat Solide, Masson, Paris: 1-11 (1972); Joets A. et Ribotta R., *J. Physique,* 47: 595-606 (1986).
15. Dreyfus J.-M. and Guyon E., *J. Physique,* **42**: 283-292 (1981).

16. Dubois-Violette E., Durand G., Guyon E., Manneville P. and Pieranski P., *Solid State Physics,* Suppl. **14**: 147 (1978).

17. Bouligand Y. in *Liquid Crystalline Order in Polymers,* Blumstein ed., Acad. Pr., N.-Y.: 261-297 (1978) and Comptes Rendus Acad. Sci., Série Générale (La Vie des Sciences), **2**: 121-140 (1985).

18. Livolant F., *Eur. J. Cell Biol.,* **33**: 300 (1984).

19. Bouligand Y., *J. Physique,* Paris, **36**: C1-173 (1974).

20. Bouligand Y. in *Dislocations in Solids,* F.R.N. Nabarro ed., North Holl., 5: 299 (1980) and in *Physics of Defects,* R. Balian *et al.,* ed., Ecole de Physique Théorique des Houches, 35, 665 North Holl. (1981).

STABILITY ANALYSIS OF FLOWS OF LIQUID CRYSTAL POLYMERS EXHIBITING A BANDING STRUCTURE

MARIA-CARME CALDERER*

Abstract. A prototype model to describe flows of *liquid crystal polymers* is analyzed. Under appropriate hypotheses on the constitutive functions, we show existence of steady-state solutions presenting a banding structure parallel to the shear direction, in shear flow geometry. A stability analysis of such solutions is carried out.

0. Introduction. In this article we discuss some results concerning shear flows of *uniaxial nematic liquid crystal polymers.* We consider a prototype model due to Ericksen [5]. We assume that liquid crystal polymers are incompressible anisotropic fluids. Accordingly, in addition to the velocity field $\mathbf{u}$ and the hydrostatic pressure p, the variables to describe flows of such materials include the director $\mathbf{n}$ and the order parameter s, both associated with the anisotropy of the material. s measures the quality of the molecular alignment along the director. ($s = 0$ corresponds to the isotropic state, with randomly oriented molecules, $s = 1$ to perfect alignment and $s = -\frac{1}{2}$ to having the molecules placed on a plane perpendicular to the director)[6]. We ignore thermal effects as well as the presence of external forces and electromagnetic fields.

Under appropriate hypotheses on the constitutive functions, we show that the governing equations posses a multiplicity of steady state solutions, which we regard as distinct nematic phases of the same material. We examine under what conditions distinct phases may coexist in the same flow and discuss the stability of such *multiple -phase solutions*. Across the interface separeting two distinct phases $\mathbf{n}$ and s become discontinuous.

The present model shares some of the features of the Leslie-Ericksen equations for ordinary liquid crystals. In the latter case, s is assumed to be constant; the flow affects the molecular orientation by causing the director to vary. Polymers exhibit higher molecular weight and therefore, the interaction flow orientation is more complex. Accordingly the present model allows for variations of the order parameter in space and time. On the other hand, it differs from the theory propossed by Doi [4] which does not allow spatial variations of the order parameter.

Liquid crystal polymers may develop textures during the flow These are regions of the flow with uniform values of the orientation and order parameter, which become discontinuous across the boundary. The present model might be appropriate to describe textures because it allows for a multiplicity of steady state solutions corresponding to a given velocity gradient. Such feature, which is not exhibited by the Leslie-Ericksen equations, is due to the presence of the order parameter in the model.

---

*Department of Mathematics, George Mason University , Fairfax, Virginia 22030. This research was supported by grant no. DMS-8714152 from the National Science Foundation.

The issue of stability of textures from experimental point of view does not seem to be completely resolved. Wissbrun [10] states that such textures in liquid crystal polymers, may be very persistent: starting with a uniform texture-free sample, texture may be developed by flow, and once formed, may then persist indefinitely. On the other hand, Berry [2] describes experimental results of texture samples evolving very slowly into uniform states.

The stress coefficients appearing in the constitutive equation for the stress tensor are functions of s determined from molecular theory arguments. Other basic assumptions on the constitutive equations are made so as to match available experimental results. In particular, the free energy is taken so that the governing equations give multiple orienting nematic phases in shear flow. This corresponds to polymers with high concentration values, above the isotropic range.

Liquid crystal polymers present very high Miesowitcz viscosities, as compared with ordinary liquid crystals. Consequently, the time -independent linear system obtained by linearizing the governing equations about shear flow steady states is elliptic. Likewise, for sufficiently large Miesowitcz viscosities the bilinear form associated with the weak formulation of the problem is coercive. However, both properties still hold under more general conditions, which for a given steady state, can be formulated in terms of the viscosity coefficients. Furthermore, the coercivity condition also involves the speed of the shear flow. Such fact might be in agreement with the experimental evidence of dependence of the Miesowitcz viscosity on the geometry of the apparatus [9].

We consider shear flows on semi-infinite strips and study solutions of the linear, second order system of integro-differential equations given in section 4, subject to the boundary conditions (5.3) and (5.4). Sincethe domain exhibits corner singularities, the velocity fields at the boundary are required to satisfy compatibility conditions so as to obtain regularity of the solutions[7]. In sections 5 and 6, we obtain local stability results for single and multiple-phase flows, respectively. In particular, when distinct phases are present in the flow, such results suggest that a banding pattern of finite length may persist, under time-independent, non-homogeneous perturbations. Consequently, it poses restrictions on the width of the bands, for a given number of them present in the flow, or viceversa, it restricts the number of bands of a given width.

Additional conditions on the elastic stress are required in order to obtain results on global stability. Such conditions turn out to be necessary and sufficient to guarantee stability under time-dependent, homogeneous perturbations.

1. Ericksen's Equations for flows of Liquid Crystal Polymers.

In order to describe time-independent, plane flows of liquid crystal polymers in a region $\Omega \subset R^2$, using the model proposed by Ericksen [5], we need to introduce the following vector fields

$$\mathbf{u} = (u(\mathbf{x}), v(\mathbf{x})) \tag{1.1}$$

$$\mathbf{n} = (n_1(\mathbf{x}), n_2(\mathbf{x})), \tag{1.2}$$

that represent the velocity of the fluid and the director, respectively, at a point $\mathbf{x} \in \Omega$. Moreover, we let the scalar field

$$s = s(\mathbf{x}) \tag{1.4}$$

Since we assume that the fluid is incompressible

$$\nabla \cdot \mathbf{u} = 0 \tag{1.5}$$

is supossed to hold for all $\mathbf{x} \in \Omega$.

The equation of balance of linear momentum as well as the evolution equations for $\mathbf{n}$ and s are given by

$$\rho\dot{\mathbf{u}} = \nabla \cdot \mathbf{t} \tag{1.6}$$

$$\gamma_1(s)\mathbf{n} \times \dot{\mathbf{n}} = \gamma_1(s)\mathbf{n} \times \omega\mathbf{n}\mathbf{n} - \gamma_2(s)\mathbf{n} \times d\mathbf{n} \tag{1.7}$$

$$2\beta_2(s)\dot{s} = -\Phi'(s) - 2\beta_1(s)\mathbf{n} \cdot d\mathbf{n} \tag{1.8}$$

$$d = \frac{1}{2}(\nabla\mathbf{u} + \nabla\mathbf{u}^T) \qquad \omega = \frac{1}{2}(\nabla\mathbf{u} - \nabla\mathbf{u}^T) \tag{1.9}$$

(The superimposed dots denote material time-derivatives). The constitutive equation for the Cauchy stress tensor, in the present model, is as follows

(1.10.)

$$\mathbf{t} = -pI + \alpha(s)d + \alpha_2(s)(\mathbf{n} \otimes d\mathbf{n} + d\mathbf{n} \otimes \mathbf{n}) + 2\alpha_1(s)(\mathbf{n} \cdot d\mathbf{n})\mathbf{n} \otimes \mathbf{n} +$$
$$\gamma_2(\dot{\mathbf{n}} \otimes \mathbf{n} + \mathbf{n} \otimes \dot{\mathbf{n}} - \omega\mathbf{n} \otimes \mathbf{n} - \mathbf{n} \otimes \omega\mathbf{n}) + 2\beta_1\dot{s}\mathbf{n} \otimes \mathbf{n}$$

p denotes the hydrostatic pressure and $\alpha, \alpha_1, \alpha_2, \beta_1, \beta_2, \gamma_1, \gamma_2$ denote constitutive functions of s, *the stress coefficients,* whose form is obtained from molecular theory argumetns. $\Phi(s)$ represents the Helmholtz free energy.

Since we want to study plane flows, we let

$$\mathbf{n} = (\cos\theta, \sin\theta), \qquad \theta \in (0, \pi) \tag{1.11}$$

In order to study the system (1.6)-(1.11) the following combination of viscosity coefficients becomes relevant:

$$\beta \equiv \alpha_2 - \frac{{\gamma_2}^2}{\gamma_1} \qquad \gamma \equiv 2(\alpha_1 + \frac{{\gamma_2}^2}{\gamma_1} - \frac{{\beta_1}^2}{\beta_2}) \tag{1.12}$$

Necessary and sufficient conditions for the constitutive equations to satisfy the Clausius-Duhem inequality are expressed in terms of relations involving the stress coefficients. For $s \in (-\frac{1}{2}, 1)$ fixed, such inequalities are anlogous to those for small molecule liquid crystals[9], namely,

$$\alpha(s) > 0, \tag{1.13}$$

$$\alpha(s) + \beta(s) > 0 \tag{1.14}$$

$$\alpha(s) + \beta(s) + \frac{2}{3}\gamma(s) + \frac{1}{3}\beta(s) > 0 \tag{1.15}$$

In order to interpret the previous inequalities, it is useful to refer to the anisotropic viscosity of a nematic liquid crystal as determined by Miesowicz. He oriented a sample by applying a magnetic field and measured the viscosity coefficients in the following three geometries[9]:

(*i*) **n** parallel to the flow.

(*ii*) **n** parallel to the velocity gradient.

(*iii*) **n** perpendicular to the flow and the velocity gradient.

The apparent viscosity for any geometry is $\eta_M = \frac{t_{ij}}{2d_{ij}}$ For a plane shear flow, in cases (i) and (ii), it gives the common value $\eta_M = \frac{1}{2}(\alpha + \beta)$, whereas the corresponding one to (iii) is identically zero. Therefore, inequality (1.14) establishes the positivity of the Miesowicz viscosity, whereas (1.13) that of the newtonian one.

2.Homogeneous plane shear flow solutions. We consider solutions to the governing equations satisfying

$$u_y = \kappa, \qquad v = 0, \qquad \theta = \theta^0, \qquad s = s^0 \tag{2.1}$$

where $\kappa \geq 0$, θ^0, $s^0 > 0$ are constant. θ^0 gives the angle of molecular alignment with the flow. Such constants are obtained as solutions of the algebraic equations

$$\gamma_1(s) + \gamma_2(s)\cos 2\theta = 0 \tag{2.2}$$

$$\Phi'(s) + \kappa\beta_1(s)\sin 2\theta = 0 \tag{2.3}$$

We let $X_0 = (\kappa, \theta^0, s^0)$ denote a steady-state shear flow solution. We assume that $\gamma_1, \gamma_2, \beta_1, \beta_2$ and Φ are smooth functions of s such that

(H1) $\gamma_1(s) = O(s^2)$, $\gamma_2(s) = O(s)$, $\beta_1(s) = O(1)$, $\beta_2(s) = O(1)$, as $s \to 0$

(H2) $\Phi'(0) = 0$, $\Phi'(s) < 0$, for $s < 0$

(H3) $\lim_{s \to 1} \Phi'(s) = \infty$ and $\lim_{s \to -\frac{1}{2}} \Phi'(s) = -\infty$

Additional hypotheses are required to ensure the solvability of equations (2.2) and (2.3), according to the experimental observation of a multiplicity of nematic states in shear flow. Such states correspond to distinct values of s in the interval $0 < s < 1$ and both, positive and negative angles of molecular orientation.

$$(H4)\quad \text{There exist constants} \quad 0 < s_M < s_m < 1 \quad \text{such that}$$
$$0 < \Phi'(s_M) \equiv M_M \quad \text{is a local maximum}$$
$$0 > \Phi'(s_m) \equiv M_m \quad \text{is a local minimum}$$

$$(H5)\quad \beta_1(s) \leq 0, \qquad |\frac{\gamma_1(s)}{\gamma_2(s)}| \leq 1, \qquad \text{for} s \in (0,1).$$

Let $s_1, s_2 > 0$ be such that $\Phi'(s_i) = 0, \quad i = 1, 2.$ It follows from equations (2.2) and (2.3) that

$$\cos 2\theta = -\frac{\gamma_1(s)}{\gamma_2(s)} \tag{2.4}$$

$$\Phi'(s) \pm \kappa h(s) = 0, \quad \text{where} \tag{2.5 a$_+$,a$_-$}$$

$$h(s) \equiv \beta_1(s)(1 - \frac{\gamma_1(s)^2}{\gamma_2(s)^2})^{\frac{1}{2}} \tag{2.6}$$

hold for $s \neq 0$.

The solvability of such equations is ensured by hypotheses (H1)-(H5) and can be summarized as follows:

(i) For every $\kappa \geq 0$ equation (2.5 a$_+$) has at least one nonzero solution $\tilde{s}(\kappa)$. Moroever it has two positive solutions, $\tilde{s}_i(\kappa), \quad i = 1, 2$ provided that κ satisfies the inequality

$$0 \leq \kappa < \kappa_M, \qquad \text{with} \quad \kappa_M \equiv \Phi'(s_M)h(s_M)^{-1} \tag{2.7}$$

(ii) For

$$0 \leq \kappa < \kappa_m, \qquad \text{with} \quad \kappa_m \equiv |\Phi'(s_m)h(s_m)^{-1}| \tag{2.8}$$

equation (2.5 a$_-$) has at least two positive solutions $\hat{s}_i(\kappa), \quad i = 1, 2$.

Whithout loss of generality, we assume that s_m and s_M are the only local extrema of $\Phi'(s), \quad s \in (0, 1)$. Therefore solutions of (2.5 a$_+$), in the interval $[s_2, 1)$, define a function

$$s = \tilde{s}(\kappa), \qquad \kappa \geq 0, \qquad \text{such that} \quad \tilde{s}(0) = s_2 \tag{2.9}$$

Likewise, solutions of (2.5 a$_-$), in the interval $[s_1, s_m]$, also define a function

$$s = \hat{s}(\kappa), \qquad \kappa \in [0, \kappa_m] \quad \text{such that} \quad \hat{s}(0) = s_1, \quad \hat{s}(\kappa_m) = s_m. \tag{2.10}$$

For $\kappa \geq 0$ fixed, we may regard the solutions (2.9) and (2.10), together with their corresponding orientation values, as two different phases of the same material. Moreover we observe that if κ satisfies inequalities (2.7) and (2.8) four distinct nematic phases are possible.

Remark. The special values s_1, s_2, s_M and s_m are completely determined by precribing the function $\Phi'(s)$, $s \in (0,1)$. Likewise, the positive numbers κ_m and $\tilde{s}_m$ depend only on the function $h^{-1}(s)\Phi'(s)$, $s \in (0,1)$.

The following notation will be useful in the forhcomming sections:

$$\beta \equiv \alpha_2 - \frac{{\gamma_2}^2}{\gamma_1}, \qquad \gamma \equiv 2(\alpha_1 + \frac{{\gamma_2}^2}{\gamma_1} - \frac{{\beta_1}^2}{\beta_2}) \tag{2.11}$$

3. Texture patterns parallel to the shear direction. We seek additional conditions on the constitutive functions in order to prove existence of steady-state flows with two or more coexisting phases. Specifically, we want to rearrange the solutions discussed in the previous section as bands parallel to the shear direction. Across the interface separating two neighboring bands, s and θ may be discontinuous. In addition to the hypotheses previoustly established, we will assume the following

(H6) $\alpha(s), \alpha_1(s), \alpha_2(s), \beta_1(s), \beta_2(s), \gamma_1(s)$ and $\gamma_2(s)$ are polynomials

We let

$$\alpha(s) \equiv (A_1 + A_2 s + A_3 s^2) \tag{3.1}$$

It follows as a consequence of inequlity (1.13) that

$$A_1 \neq 0 \tag{3.2}$$

Remark. Not all the coefficients of such polynomials are independent. The positivity requirement on the dissipation function as well as arguments from molecular theory stablish certain relationships among the cofficients of α, α_1 and α_2 as well as among those of $\gamma_1, \gamma_2, \beta_1$ and β_2 [5].

The shear rate κ must be the same in two neighboring bands, in order to achieve balance of mass across the interface. Likewise, balance of linear momentum implies continuity of the stress components t_{11} and t_{12}. The continuity of the first one can be achieved by adjusting p in both regions accordingly. We then consider a steady-state shear flow with shear rate $\kappa \in [0, \kappa_m]$ such that

$$s = \tilde{s}(\kappa) \qquad \sin\theta^0 = [\frac{1}{2}(1 + \frac{\gamma_1}{\gamma_2}(s)]^{\frac{1}{2}} \text{ for } x > x_0 \tag{3.3a+}$$

$$s = \hat{s}(\kappa) \qquad \sin\theta^0 = -[\frac{1}{2}(1 + \frac{\gamma_1}{\gamma_2}(s)]^{\frac{1}{2}} \text{ for } x < x_0, \tag{3.3a-}$$

where $x_0 \in R$ describes the interface, with κ determined so that t_{12} is continuous. Equivalently, κ must be obtained as a solution of the equation

(3.4) $$F(\kappa) = 0, \quad \text{where}$$

(3.5) $$F(\kappa) = \kappa[H(\tilde{s}(\kappa)) - H(\hat{s}(\kappa))],$$

(3.6) $$H(s) \equiv \alpha(s) + \beta(s) + (\gamma(s) + \frac{{\beta_1}^2(s)}{\beta_2(s)})(1 - \frac{{\gamma_1}^2(s)}{{\gamma_2}^2(s)})$$

The following is a sufficient condition to ensure the solvability of equation (3.4):

(3.7) $$F'(0)F(\kappa_m) < 0$$

We call solution pairs of equations (3.3)-(3.6), compatible steady-states. It follows from (1.12), (3.1)- (3.4) and the remark at the end of section 2 that $F(\kappa)$ does not involve the parameter A_1. Consequently, inequality (3.7) provides a single relationship among the remaining parameters in the viscosity coefficients of (1.10), for every $s \in (0,1)$. For instance, such an inequality can be satisfied, for $\beta_1, \beta_2, \gamma_1$ and γ_2 prescribed, by appropriately choosing the range of one of the parameters in α, α_1 and α_2. In particular, we may regard inequality 3.7 as an additional restriction on the parameter A_2.

The previous observation leads to the following conclusion: for a given set of constitutive functions satisfying (H1)-(H6) (3.1), (3.2) and (3.7), there exists at least one value of κ in the interval $(0, \kappa_m)$ that solves equation (3.4). To such value of κ, there corresponds a flow of the form (3.3). Moreover, the pair of compatible steady states can be alternatively matched to form banding structures, with s and θ becomming discontinuous across interfaces. In particular, periodic arrangements are also allowed. Equation (3.4) is also satisfied for $\kappa = 0$ and order parameter values s_1 and s_2, for $x < x_0$ and $x > x_0$, respectively. Such situation corresponds to either equilibrium or uniform flow.

4. Linear stability analysis of steady-state shear flows. We let $\mathbf{u}, p, \theta, s$ denote perturbations of a steady-state shear-flow X_0, as in (2.1). The linearization of the governing equtions (1.5)-(1.11) about such particular solution yields the following(linear) system

(4.1)
$$\rho\mathbf{u}^0.\nabla\mathbf{u} = -p_x + \mathcal{A}_1 u_{xx} + \mathcal{B}(u_{xy} + v_{xx}) + \mathcal{C}(u_{yy} + v_{xy}) + \mathcal{E}_1 v_{yy} + \mathbf{m}_1.\nabla\theta + \mathbf{n}_1.\nabla s - \rho\kappa v,$$
(4.2)
$$\rho\mathbf{u}^0.\nabla v = -p_y + \mathcal{B}u_{xx} + \mathcal{C}(u_{xy} + v_{xx}) + \mathcal{E}_2(u_{yy} + v_{xy}) + \mathcal{A}_2 v_{yy} + \mathbf{m}_2.\nabla\theta + \mathbf{n}_2.\nabla s,$$
(4.3) $$\mathbf{u}^0.\nabla\theta = m_3\theta + n_3 s + a_3 u_x + b_3 u_y + c_3 v_x + d_3 v_y,$$
(4.4) $$\mathbf{u}^0.\nabla s = m_4\theta + n_4 s + a_4 u_x + b_4 u_y + c_4 v_x + d_4 v_y$$
(4.5) $$0 = u_x + v_y,$$

where

$$(4.6)\quad \mathcal{A}_1 = \alpha^0 + (2\beta^0 + \gamma^0 \cos 2\theta^0)\cos^2\theta^0 \qquad \mathcal{A}_2 = \alpha^0 + (2\beta^0 - \gamma^0 \cos 2\theta_0)\sin^2\theta^0$$

$$(4.7)\quad \mathcal{B} = \frac{1}{2}\sin 2\theta^0(\beta^0 + \gamma^0\cos^2\theta^0), \qquad \mathcal{C} = \frac{1}{2}(\alpha^0 + \beta^0 + \frac{\gamma^0}{2}\sin^2 2\theta^0)$$

$$(4.8)\quad \mathcal{E}_1 = -\frac{\gamma^0}{4}\sin 4\theta^0, \qquad \mathcal{E}_2 = \frac{1}{2}\sin 2\theta^0(\beta^0 + \gamma^0\sin^2\theta^0)$$

The zero superindices denote quantities evaluated at the given steady-state. (The specific form of the remaining coefficients is given in [3]).

It follows as a consequence of inequalities (1.13)-(1.15) that

$$(4.9)\qquad \mathcal{C} > 0$$

holds for all $s \in (-\frac{1}{2}, 1)$.

In order to analyze the system (4.1)-(4.5), we first integrate equations (4.3) and (4.4) for θ and s, with respect to x, using the variation of constants formula, with y plying the role of a parameter. We let λ_1 and λ_2 denote the eigenvalues of the corresponding non-homogeneous system. Subsequent substitution of the resulting expressions for θ and s into equations (4.1) and (4.2) gives a (linear) system of integro-differential equations fomally written as

$$(4.10)\qquad -L\mathbf{u} + \nabla p = \mathbf{F}$$

$$(4.11,)\qquad \nabla.\mathbf{u} = 0$$

where

$$(4.12)\qquad L\mathbf{u} = \begin{bmatrix} L_1^1 u + L_2^1 v \\ L_1^2 u + L_2^2 v \end{bmatrix}$$

L denotes a second order linear operator with higher order derivatives as in equations (4.1) and (4.2), whereas $\mathbf{F}$ represents an integral operator depending linearly on $\mathbf{u}$ and $\nabla\mathbf{u}$. (The kernel of such operator is of Volterra type of the second kind and consists of exponentials $e^{\lambda_i x}, i = 1, 2$). Typically , components of the operator L have the form,

$$L_1^1 u = \mathcal{A}_1 u_{xx} + \mathcal{B} u_{xy} + \mathcal{C} u_{yy} + \frac{m_1^1}{u^0}(a_3 u_x + b_3 u_y) + \frac{n_1^1}{u^0}(a_4 u_x + b_4 u_y) - \rho u^0 u_x.$$

In what follows, we shall assume that

$$(\mathrm{H7})\qquad u^0(y) \neq 0, \qquad \text{for} \quad (x, y) \in \Omega,$$

where Ω denotes the flow region. This prevents the system from becoming singular, in which case a separate treatment is required. In a shear flow such condition is achieved by appropriately prescribing the speed of both boundary plates. Alternatively, we may ignore the hypothesis (H7) and restrict our analysis to regions of the flow with non-vanishing velocity field.

5.Elliptic boundary value problem associated with shear flow regimes. We will analyze the system (4.10)-(4.12) in the case that the operator L is elliptic. This is equivalent to requiring the following inequality to hold:

$$(5.1) \qquad 1-\frac{\gamma^0}{4\mathcal{C}}[\sin^2 4\theta^0 - \cos 4\theta^0(1-\sigma\cos 4\theta^0)] \geq 0, \qquad \text{with } \sigma \equiv \frac{\gamma^0\mathcal{C}^{-1}}{|\gamma^0\mathcal{C}^{-1}|}$$

A sufficient condition for (5.1) to hold is that the ratio $\frac{|\gamma^0|}{\alpha^0+\beta^0}$ be sufficiently small.

We assume that the domain of the flow is a semi-infinite strip of width $h > 0$, i.e.

$$(5.2) \qquad \Omega = (0,\infty)\times(0,h)$$

According to (H7), we let $u_0(y) = \kappa y + C_0, \quad y \in [0,h],$ where $C_0 > 0$ denotes the speed of the lower plate.

We supplement the system (4.11)-(4.13) with the following boundary conditions,

$$(5.3) \qquad \mathbf{u}(x,0) = \hat{\mathbf{u}}^0(x), \quad \mathbf{u}(x,h) = \hat{\mathbf{u}}^h(x), \quad x \in (0,\infty)$$

$$(5.4) \qquad \mathbf{u}(0,y) = 0, \quad \theta(0,y) = \hat{\theta}^0(y), \quad s(0,y) = \hat{s}^0(y), \quad y \in [0,h]$$

We suppose that $\hat{\mathbf{u}}^0, \hat{\mathbf{u}}^h$ satisfy the following compatibility conditions:

$$(5.5) \qquad \hat{a}_1^2 \equiv \int_0^1 \frac{1}{x}\left|\frac{\partial \hat{u}^0}{\partial x}\right|^2 dx < \infty, \qquad \hat{a}_2^2 \equiv \int_0^1 \frac{1}{x}\left|\frac{\partial \hat{u}^h}{\partial x}\right|^2 dx < \infty$$

$$\int_0^\infty (\hat{v}^h(x,h) - \hat{v}^0(x,0))\,dx = 0.$$

Such conditions guarantee the extension of the velocity field on the boundary to a smooth divergence free vector field $\mathbf{a}$ in the interior [1][8]. Conditions that ensure the coercivity of the bilinear form associated with the operator L, for a given steady-state, are determined in [3]. Specifically, for $C_0 > 0$ fixed, the inequality

$$(\mathrm{H8}) \qquad \mathcal{C} > \frac{1}{k^2}\left(\frac{a_\alpha}{C_0^2} + \frac{a_\beta}{C_0} + a_\gamma + a_\delta C_0\right)$$

holds, where $a_\alpha, a_\beta, a_\gamma, a_\delta$ depend on θ_0 and s_0 and k is the constant in Korn's inequality. We observe that such hypothesis holds, in particular, for sufficiently large values of the Miesowicz viscosity.

For $R > 0$ fixed, we let $\Omega_R = (0,R)\times(0,h)$.

THEOREM 1. *[3]. Let $h > 0$ and $r > 0$ be fixed. Suppose that hypotheses (H1)-(H8),(3.1) and (3.2) hold and let $\hat{\theta}^0, \hat{s}^0 \in C^1[0,h], \hat{\mathbf{u}}^0, \hat{\mathbf{u}}^h \in [H^{\frac{3}{2}}(0,\infty)]^2$ satisfy the compatibility conditions (5.5) together with the inequality*

$$h\max(|\hat{\theta}_0|_1, |\hat{s}_0|_1) + ||\hat{\mathbf{u}}^0||_{[H^{\frac{3}{2}}(0,\infty)]^2} + ||\hat{\mathbf{u}}^h||_{[H^{\frac{3}{2}}(0,\infty)]^2} + \hat{a}_1 + \hat{a}_2 < \frac{r}{2(\alpha_0+\beta_0)}$$

Then there exists $R > 0$, $R = R(\kappa, \theta^0, s^0, C_0, h)$ such that the boundary value problem (4.1)-(4.5), (4.16)-(4.17) has a unique solution $\mathbf{u} \in [H^2(\Omega_R)]^2$, $p \in H^1(\Omega_R)$, $\mathbf{u}-\mathbf{a} \in [H_0^1(\Omega_R)]^2, \theta, s \in L^2(\Omega_R)$ such that

$$||\mathbf{u}||_{[H^2(\Omega_R)]^2} < r, \qquad ||p||_{H^1(\Omega_R)} < (\alpha^0+\beta^0)r$$

Moreover θ, s are absolutely continuous with respect to x and $\theta_x, s_x \in L^2(\Omega_R)$.

Remarks. (i) Analogous estimates for θ, s, θ_x, s_x in the L^2-norm can also be obtained, with the right hand side proportional to r. Such set of estimates establishes continuous dependence of the solution with respect to the boundary data. (ii) Global existence of solutions in Ω can be obtained under two additional assumption: $\Re\lambda_i < 0$, $i = 1,2$ and $C_0(\alpha^0+\beta^0)$ sufficiently large. (The first condition is interpreted in [3] in terms of the elastic stress $\Phi'(s)$).

6. Stability of multiphase flows. We study the stability of a stedy-state solution X_0 such that

(6.1) $$X_0 = X_0^-(\kappa) \quad \text{for} \quad x < x_0 \quad \text{and} \quad X_0 = X_0^+(\kappa) \quad \text{for} \quad x > x_0,$$

where $\kappa > 0$ is a solution of equation (3.4) and $x_0 = 0$ describes the interface separating the given pair of compatible states.

We let

(6.2) $$\Omega_1 = \{(x,y) \in \Omega : x < x_0\} \quad \text{and} \quad \Omega_2 = \{(x,y) \in \Omega : x > x_0\}, \quad \text{and}$$

(6.3) $$[f(x)] \equiv \lim_{x\to x_0^+} f(x) - \lim_{x \to x_0^-} f(x),$$

provided such limits exist. Next theorem refers to the boundary value problem, (4.1)-(4.5), (5.3)-(5.5), where the coefficients of the system belong to $L^\infty(\Omega)$ (They are those corresponding to X_0^- in Ω_1 and to X_0^+ in Ω_2).

THEOREM 2. *[3]. Let $r > 0$ be given and let κ be a solution of equation (3.2) and suppose that the hypotheses of theorem 1 as well as inequality (3.7) hold for X_0^+, X_0^-. Then* *there exists $N_0 = N_0(\kappa, h, C_0) > 0$ such that, for $x_0 < N_0$ the boundary value problem previously described has a unique solution $\mathbf{u}(x,y), p(x,y), \theta(x,y), s(x,y)$ such that,*

$$\text{(6.4)} \qquad \mathbf{u} \in [H^2(\Omega_1 \cup \Omega_{2,N_0})]^2 \qquad \mathbf{u} - \mathbf{a} \in [H_0^1(\Omega_{N_0})]^2$$

$$\text{(6.5)} \qquad p \in H^1(\Omega_1 \cup \Omega_{2,N_0}) \cap L^2(\Omega_{N_0})$$

$$\text{(6.6)} \qquad J\mathbf{u} = 0,$$

where

$$\Omega_{N_0} \equiv (x,y) \in R^2 : 0 < x < N_0, \quad 0 < y < h, \qquad \Omega_{2,N_0} \equiv \Omega_2 \cap \Omega_{N_0}$$

Moreover $\theta, s \in L^2(\Omega_{N_0})$ and their corresponding restrictions to $\Omega_i, i = 1,2,$ respectively, give absolutely continuous functions of x. Furthermore, the solution functions satisy estimates analogous to those stated in theorem 5.1, with Ω_R replaced with $\Omega_1 \cup \Omega_{2,N_0}$.□

We observe that the quantity N_0 places a restriction on the width of the bands coexisting on the interval $[0, N_0]$, or equivalently the number of alternating bands present in the flow can be increased, by reducing their correspondding widths accordingly. Equation 6.6 corresponds to the linearization of the balance of linear momentum across the interface[3]. In particular, if $\mathbf{u}$ is smooth, it has the following representation:

$$J\mathbf{u} = \{[\mathcal{A}_1 u_x + \mathcal{B}(u_y + v_x) + \frac{u}{u_0}(m_1^1 a_3 + n_1^1 a_4) + \frac{v}{u^0}(m_1^1 c_3 + n_1^1 c_4) - p],$$
$$[\mathcal{C}(u_y + v_x) + \mathcal{E} - 2v_y + \frac{u}{u^0}(m_2^1 a_3 + n_2^1 a_4) + \frac{v}{u^0}(m_2^1 c_3 + n_2^1 c_4)]\}$$

Global existence of solutions in Ω can be obtained under the two additional conditions described at the end of section 5. However, in the present case, such conditions should be satisfied by the corresponding quantities evaluated at each one of the steady states X_0^+ and X_0^-, respectively.

Conclusion. We have carried out the present analysis under general qualitative hypotheses on the constitutive functions, so as to allow including additional information from experimental measuraments, if available. For instance, no assumptions are made on the stress coefficients near $s = 1$. Instaed, they contain free parameters that could be chosen so as to match experimental data corresponding to such values of s[2]. This corresponds to the point of view of studying the main predictions of the model in the analysis of textures, rather than a more detailed information on the solutions, perhaps difficult to relate with experiments. Furthermore, the present model is in fact, a special case of a more general one exhibiting a dependence of the constitutive functions on ∇s and $\nabla \mathbf{n}$.

The present results do not cover flows of materials with concentration values near the isotropic region. Oscillatory behavior about such state may be related to texture patterns with alternating values of the director angle of ± 45 degrees. This is a subject of a separate study.

Bibliography

[1] D.N. Arnold, L.R.Scott, M.Vogelius, *Regular Inversion of the Divergence Operator with Dirichlet Boundary Conditions on a Polygon*, Annali Scuola Normale, Pisa (to appear) (1988).

[2] G.C. Berry, *Rheological Properties of Nematic Solutions of Rodlike Polymers*, Mol. Cryst. Liq. Cryst. (1988).

[3] M.C.Calderer, *Multiphase flows of liquid crystal polymers*, submitted to the Journal of Differential Equations (1988).

[4] M.Doi, *Viscoelasticity of Concentrated Solutions of Stiff Polymers*, Faraday Symp. Chem. Soc., 18 (1983), pp. 49-56.

[5] J.L.Ericksen, *On equations of motion for liquid crystal polymers*, (to appear).

[6] J.L.Ericksen, *Orienting Polymers*, J.L.Ericksen (ed.), Springer-Verlag New York, 1984, p. 27.

[7] P.Grisvard, *Boundary Value Problems in Non–Smooth Domains*, Lectures Notes no.19, University of Maryland, College Park, 1980.

[8] O.A.Ladyzhenskaya, *The Mathematical Theory of Viscous Incompressible flow*, Gordon and Breach, Science Publishers Inc., New York–London, 1963.

[9] F.Leslie, *Theory of Flow Phenomena in Liquid Crystals*, Advances in Liquid Crystals (1979), G. Brown (ed.), New York–San Francisco–London.

[10] K.F.Wissbrun, *Rheolory of Rod–like Polymers in the Liquid Crystalline State*, Journal of Rheology, 25 (1981), pp. 619-662.

Acnowledgment. I would like to thank Professor J.L.Ericksen for having propossed this model to me, as well as for the numerous discussions and suggestions on the subject.

COUPLED TRANSVERSE-LONGITUDINAL SOLITONS
IN ELASTIC CRYSTALS.

S. Cadet*
Laboratoire d'Optique du Réseau Cristallin
Faculté des Sciences, 6 bd Gabriel, F21100 DIJON, FRANCE.

1- Introduction.

We present an example of multicomponent solitons in elastic crystals. They are non linear shear waves coupled with longitudinal displacement of the atomic planes. Shear deformation has been considered in continuous models of shape memory alloys [1,2] in order to describe the domain walls between adjacent phases. The energy of deformation yield non linear dispersive dynamical equations that admit soliton solutions, but the effect of possible longitudinal displacement has not been investigated though it may be of consequence. Indeed, the study of the dynamics of an atomic chain when transverse motion only is allowed, leads to soliton modes [3-5]. However when the particles of the chain are free to move along the longitudinal direction similar solutions are derived [6] but their stability properties are much affected. Here we report some results on coupled transverse-longitudinal waves obtained using a semi-discrete reductive perturbation method [7].

2- The model.

The model considered is a monoatomic one-dimensionnal lattice with spacing a. The atoms interact with their first nearest neighbors and we assume central forces. We denote by (X_n, Y_n, Z_n) the displacement from equilibrium of the nth atom (figure 1).

* Present address:Laboratoire de Modélisation en Mécanique, Université P.et M.Curie, tour 66, 4 place Jussieu, F75252 PARIS CEDEX 07, FRANCE.

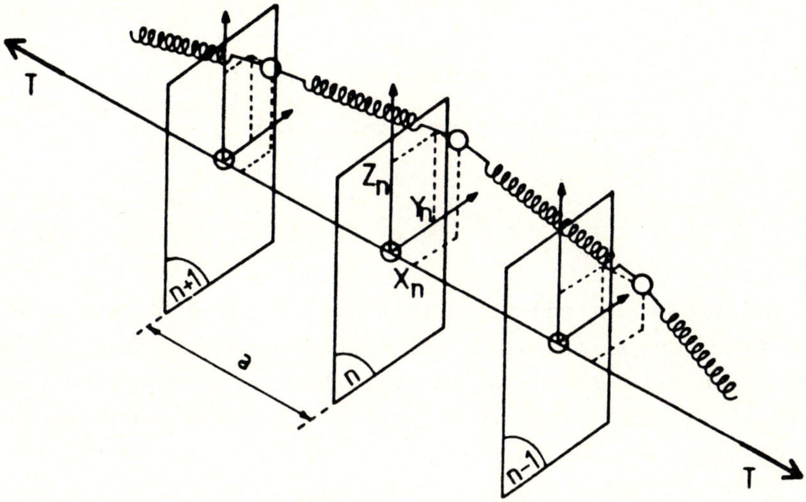

Figure 1: Model of a chain with three-dimensional displacement.

The Hamiltonian of the system is:

$$\mathcal{H} = \sum_n \frac{1}{2} m \left(\dot{X}_n^2 + \dot{Y}_n^2 + \dot{Z}_n^2\right) + \sum_n V(d_n) \tag{2.1}$$

where d_n is the bond length between particles n and (n+1):

$$d_n = ((a + X_{n+1} - X_n)^2 + (Y_{n+1} - Y_n)^2 + (Z_{n+1} - Z_n)^2)^{1/2} \tag{2.2}$$

and the interaction potential has the general polynomial form:

$$V(d) = V_1(d-a) + \frac{1}{2} V_2(d-a)^2 + \frac{1}{3!} V_3(d-a)^3 + \ldots \tag{2.3}$$

There is need to assume a permanent internal stress in the chain, that is $V_1 \neq 0$, in order to retain sufficient transverse rigidity of the chain. Indeed, if $V_1 = 0$ there is no linear term in the dynamical equations for transverse motion. Furthermore the coefficients V_1 and V_2 are chosen positive so that the equilibrium state is stable.

From the Hamiltonian we derive the equations of motion:

$$m\,\ddot{X}_n = V'(d_n)\,\frac{a + X_{n+1} - X_n}{d_n} - V'(d_{n-1})\,\frac{a + X_n - X_{n-1}}{d_{n-1}} \tag{2.4.a}$$

$$m\,\ddot{Y}_n = V'(d_n)\,\frac{Y_{n+1} - Y_n}{d_n} - V'(d_{n-1})\,\frac{Y_n - Y_{n-1}}{d_{n-1}} \tag{2.4.b}$$

$$m\,\ddot{Z}_n = V'(d_n)\,\frac{Z_{n+1} - Z_n}{d_n} - V'(d_{n-1})\,\frac{Z_n - Z_{n-1}}{d_{n-1}} \tag{2.4.c}$$

The dynamical equations (2.4) yield two different kinds of non linearities: (i) a material non linearity due to the non linear dependence of the bond force $V'(d)$ on the interatomic distance; (ii) a geometrical non linearity that results of the dependence of the bond length on the displacements as stated by (2.3) and of the ratios that come from the projections of the force along the axis of coordinate.

Since the discrete set of equations (2.4) is not tractable, we restrict ourselves to small deviations of d_n from a. Then expanding d_n with respect to the components of the relative displacements we obtain approximate equations of motion that retain the necessary terms in view of the following calculations.

$$\ddot{X}_n = \frac{C_L^2}{a^2}\,(X_{n+1} + X_{n-1} - 2X_n) + \frac{C_A}{a^3}\Big((X_{n+1} - X_n)^2 - (X_n - X_{n-1})^2\Big) + \frac{C_L^2 - C_T^2}{2a^3}\Big(|H_{n+1} - H_n|^2 - |H_n - H_{n-1}|^2\Big) \tag{2.5.a}$$

$$\begin{aligned}\ddot{H}_n = {} & \frac{C_T^2}{a^2}\,(H_{n+1} + H_{n-1} - 2H_n) \\ & + \frac{C_L^2 - C_T^2}{a^3}\Big((X_{n+1} - X_n)(H_{n+1} - H_n) - (X_n - X_{n-1})(H_n - H_{n-1})\Big) \\ & + \frac{C_L^2 - C_T^2}{2a^4}\Big(|H_{n+1} - H_n|^2(H_{n+1} - H_n) - |H_n - H_{n-1}|^2(H_n - H_{n-1})\Big)\end{aligned} \tag{2.5.b}$$

where we introduce the complex variable $H_n = Y_n + iZ_n$ that describe both components of transverse displacement, and set:

$$C_T^2 = \frac{V_1\,a}{m} \qquad C_L^2 = \frac{V_2\,a^2}{m} \qquad C_A = \frac{V_3\,a^3}{2m} \tag{2.6}$$

The coefficients C_T and C_L are respectively the velocity of sound for transverse and longitudinal linear waves with long wavelength, C_A is a coefficient arising from material nonlinearity. Since C_L is greater than C_T in simple lattices we confine our attention to this case.

Before proceeding to the presentation of solitary waves, the set of coupled difference- differential equations (2.5) deserves some comments.

If both longitudinal and transverse displacements are present and their amplitudes are sufficiently small, the non linear terms in (2-5) can be neglected. Looking for harmonic sinusoidal solutions yields the dispersion relations for longitudinal and transverse waves:

$$\omega_L^2 = 2C_L^2(1 - \cos ka)/a^2 \qquad (2.7.a)$$

$$\omega_T^2 = 2C_T^2(1 - \cos ka)/a^2 \qquad (2.7.b)$$

where ω_L, ω_T are angular frequencies and k is the wave number.

Next we note that equations (2.5) have purely longitudinal solutions, with $Y_n = Z_n = 0$. This is the situation extensively studied by previous authors (see for example reference 8), soliton solutions then represent compressive or rarefactive waves.

But, while in the linear approximation transverse and longitudinal motions are uncoupled, when the amplitude of transverse displacement is sufficiently large a coupling term in (2.5.a) forces longitudinal motion. As a consequence non linear transverse waves are always accompanied by a longitudinal deformation.

In the following we are interested in examining the longitudinal distorsion coupled with transverse envelope solitary waves.

3- Circularly polarized envelope modes.

We consider a circularly polarized transverse harmonic wave modulated by an envelope function F_n:

$$H_n = \varepsilon F_n(t).\exp i(kna - \omega t) \qquad (3.1.a)$$

$$X_n = \varepsilon G_n(t) \qquad (3.1.b)$$

where the wavenumber k and the frequency ω are related by the dispersion relation for transverse waves (2.7.b) and ε is a small scaling parameter.

Inserting (3.1) in (2.5) we get:

$$\varepsilon G_{n,tt} = \varepsilon \frac{C_L^2}{a^2}(G_{n+1} + G_{n-1} - 2G_n) + \varepsilon^2 \frac{C_A}{a^3}\Big((G_{n+1} - G_n)^2 - (G_n - G_{n-1})^2\Big)$$
$$+ \varepsilon^2 \frac{\beta}{2a}\Big(\left|F_{n+1}e^{i\theta} - F_n\right|^2 - \left|F_n - F_{n-1}e^{-i\theta}\right|^2\Big) \qquad (3.2)$$

$$-\omega^2 \varepsilon F_n - 2i\omega\varepsilon F_{n,t} + \varepsilon F_{n,tt} = \varepsilon \frac{C_T^2}{a^2}\Big(F_{n+1}e^{i\theta} + F_{n-1}e^{-i\theta} - 2F_n\Big)$$
$$+ \varepsilon^2 \frac{\beta}{a}\Big((G_{n+1} - G_n)\big(F_{n+1}e^{i\theta} - F_n\big) - (G_n - G_{n-1})\big(F_n - F_{n-1}e^{-i\theta}\big)\Big)$$
$$+\varepsilon^3 \frac{\beta}{2a^2}\Big(\left|F_{n+1}e^{i\theta} - F_n\right|^2\big(F_{n+1}e^{i\theta} - F_n\big) - \left|F_n - F_{n-1}e^{-i\theta}\right|^2\big(F_n - F_{n-1}e^{-i\theta}\big)\Big) \qquad (3.3)$$

where θ = ka and $\beta = \dfrac{C_L^2 - C_T^2}{a^2}$.

The fast oscillating exponential terms have cancelled and k occurs only in the phase constants θ, the wave number can then take any value. Now we assume the spatial variations of F_n and G_n to be slow and we introduce the stretched coordinates $\xi = \varepsilon x$, $\tau = \varepsilon t$. Going to the continuum limit, we obtain to order ε^3:

$$\varepsilon^3 G_{\tau\tau} = \varepsilon^3 C_L^2 G_{\xi\xi} + \varepsilon^3 \beta\,(1 - \cos\theta)\,\big(|F|^2\big)_\xi \qquad (3.4)$$

$$-2i\omega\varepsilon^2 F_\tau + \varepsilon^3 F_{\tau\tau} = \frac{C_T^2}{a^2}\,(2ai\varepsilon^2 \sin\theta\; F_\xi + a^2\varepsilon^3 \cos\theta\; F_{\xi\xi})$$
$$+ 2\beta(\cos\theta - 1)\varepsilon^3 F G_\xi - 2\frac{\beta}{a^2}(1 - \cos\theta)^2 \varepsilon^3 |F|^2 F \qquad (3.5)$$

We go now to a frame moving at the group velocity of the carrier wave:

$$s = \xi - v_g\tau \qquad \sigma = \varepsilon\tau \qquad v_g = \frac{d\omega}{dk} = \frac{C_T^2}{\omega a}\sin\theta$$

Equations (3.4), (3.5) then yield at order ε^3:

$$G_s = -\frac{\beta}{C_L^2 - v_g^2}(1 - \cos\theta)\,|F|^2 \tag{3.6}$$

and we obtain for F a Non Linear Schrödinger equation:

$$-iF_\tau + PF_{ss} + Q|F|^2F = 0 \tag{3.7}$$

where $P = \dfrac{v_g^2 - C_T^2\cos\theta}{2\omega}$ $\qquad Q = \dfrac{\beta}{\omega a^2}(1 - \cos\theta)^2\left(1 - \dfrac{\beta a^2}{C_L^2 - v_g^2}\right)$

Since the product PQ is positive for any k, the solutions of (3.7) are localized envelopes [9]:

$$F = \frac{1}{\sqrt{Q}}\,\mathrm{sech}\left(\frac{s}{\sqrt{2P}}\right).\, e^{-i\frac{\tau}{2}} \tag{3.8}$$

Figure 2: (a) circularly polarized envelope and (b) its coupled longitudinal displacement.

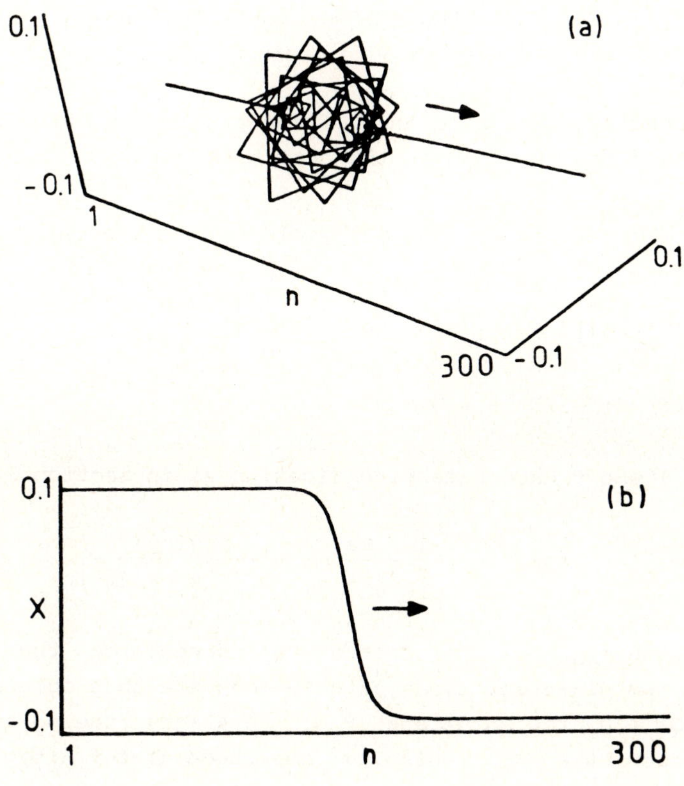

Going back to the original variables H_n, X_n, we find that a transverse circularly polarized localized envelope is coupled with longitudinal displacement in the form of a compressive kink as illustrated on figure 2. However numerical calculation show that the interatomic bonds are lengthened.

4- Linearly polarized envelope modes.

We assume that transverse motion is polarized along Y_n for example, we seek solutions of the form:

$$H_n = Y_n = \frac{\varepsilon}{2} F_n(t).\exp i(kna - \omega t) + c.c.$$

$$X_n = \varepsilon G_n(t) + \frac{\varepsilon^2}{2}\left(R_n(t).\exp 2i(kna - \omega t) + c.c.\right) \tag{4.1}$$

where c.c. stands for complex conjugate.

In (4.1) we have neglected higher harmonics which bring no contribution to the calculation of F_n, G_n and R_n.

A procedure similar to that of the previous section leads to the following equations obtained in the continuum limit:

$$\left(v_g^2 - C_L^2\right) G_{ss} = -\frac{\beta}{2}(1 - \cos\theta)\left(|F|^2\right)_s \tag{4.2}$$

$$\left(-4\omega^2 + \omega_L^2(2k)\right)R = i\,\frac{\beta}{a}\sin\theta(\cos\theta - 1)F^2 \tag{4.3}$$

$$-iF_\sigma + PF_{ss} + Q'|F|^2F = 0 \tag{4.4}$$

where s, σ, θ and P have same signification as in section 3 and:

$$Q' = \frac{\beta}{\omega a^2}(1 - \cos\theta)^2\left(\frac{3}{4} - \frac{1}{2}.\frac{\beta a^2}{C_L^2 - v_g^2} - \frac{\beta\sin^2\theta}{\omega_L^2(2k) - 4\omega^2}\right) \tag{4.5}$$

Here the sign of the product PQ' depends on the wavenumber k (figure 3), two different types of solutions are thus obtained [10]:

(i) if $PQ' < 0$ the solutions F of (4.4) are the so-called "dark" solitons which have non vanishing amplitude at infinity. This is the case when:

$$2\,\frac{C_L^2 + 2C_T^2}{2C_L^2 + C_T^2} - 1 < \cos\theta < 2\,\frac{C_T^2}{C_L^2} - 1$$

(ii) if PQ' > 0 the solutions of (4.4) are localized envelopes.

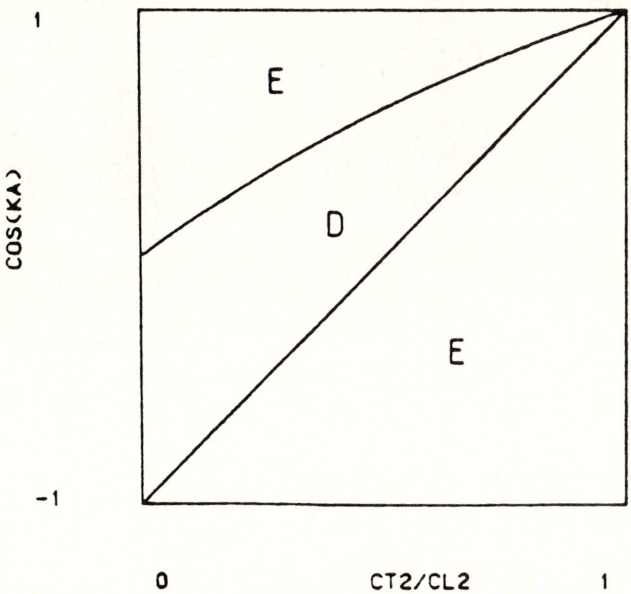

Figure 3: Plot of the curves separating the regions where envelopes (E) and dark solitons (D) are stable.

The instability of localized envelopes originates in a resonance between transverse and longitudinal oscillations when $4\omega^2$ approaches $\omega_L^2(2k)$. Indeed, the longitudinal oscillation which has amplitude R given by (4.3), can propagate independently from the transverse wave since its wave number 2k and its frequency 2ω satisfy the dispersion relation (2.7.a).

The longitudinal displacement associated with a transverse localized envelope consists in a compressive kink to which is added a smaller localized oscillation. When dark solitons are solutions of (4.4) the longitudinal displacement is the sum of a uniform compression, a rarefactive kink and a small oscillation that takes on the form of a dark wave.

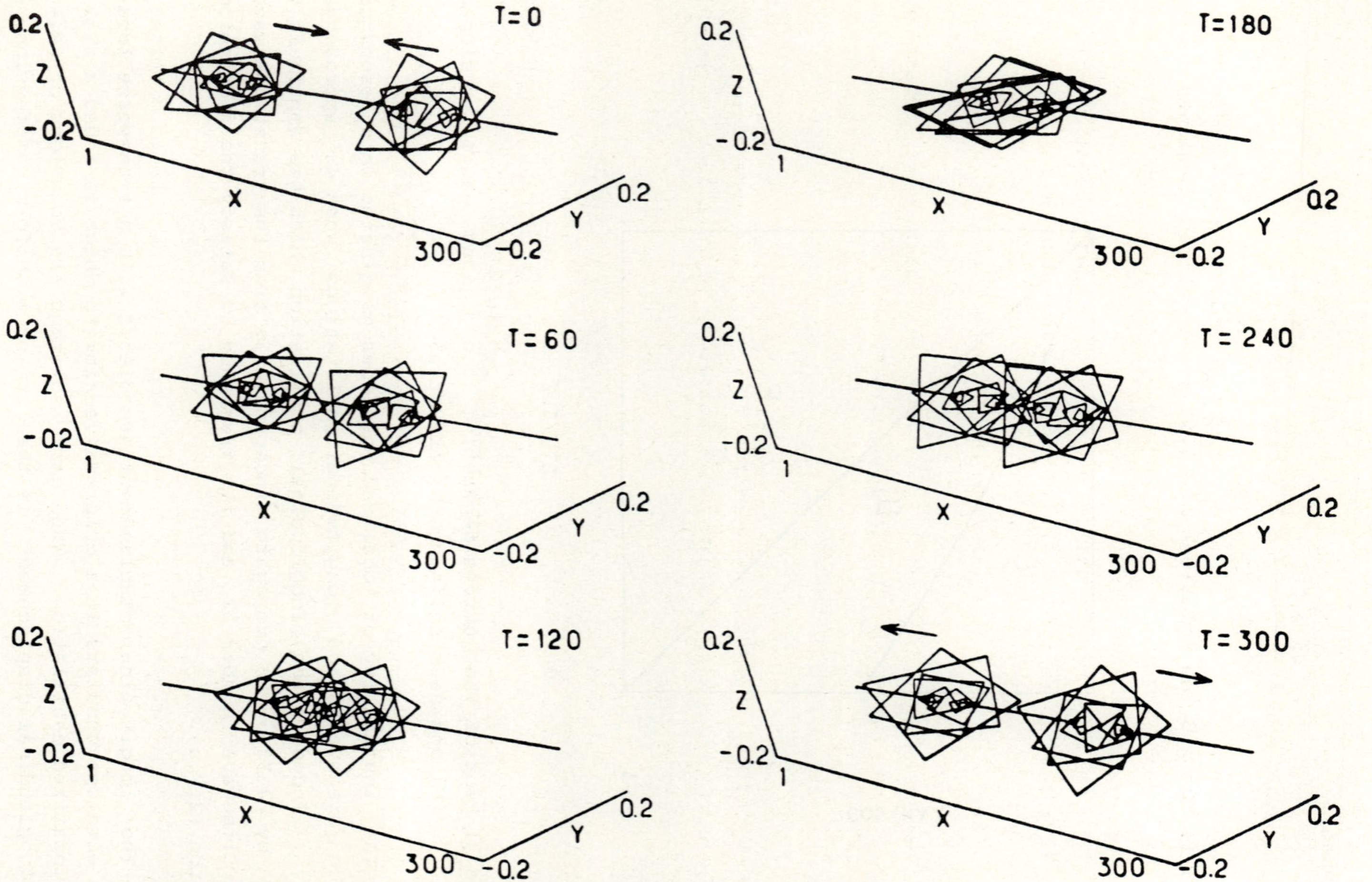

Figure 4: Collision of two circularly polarized envelopes.

5- Stability of localized envelopes.

Numerical simulations show that the solutions of the dynamical equations obtained in the semi-discrete approximation, propagate in the discrete lattice after very small adjustment of their shape.

On the other hand they are not exact solitons because the collision of two envelopes results in an important loss of energy of both waves when their frequencies and wave number are such that a longitudinal free mode is generated by frequency mixing during the interaction.

The collision of circularly polarized envelopes with parameters (ω_1, k_1) and (ω_2, k_2) is inelastic if either of the two following conditions is satisfied:

(i) $k_1 - k_2 = 0$, $\omega_1 = -\omega_2$.
Two longitudinal wave packets are emitted in opposite directions, their frequency is $\omega_L = 2\omega_1$ and the wave vector k_L is related to ω_L by the dispersion relation (2.7.b).

(ii) $\omega_L = \omega_1 - \omega_2$ and $k_L = k_1 - k_2$ satisfy the dispersion relation (2.7.b). Then one longitudinal wave packet (ω_L, k_L) is emitted.

When at least one linearly polarized wave is involved, the same conditions hold with parameters $(\pm\omega_i, \pm k_i)$.

Beside inelastic behavior, while circularly polarized envelopes undergo only a phase shift and a center shift (figure 4), a small change of shape of linearly polarized envelopes is observed after a collision with another wave. More specifically, the polarization becomes slightly elliptic, the major axis making a small angle with the initial direction of linear polarization (figure 5).

These phenomena can be discussed with the help of a refined version of the reductive perturbation method [5,11] from which phase shifts and variations of polarization are obtained.

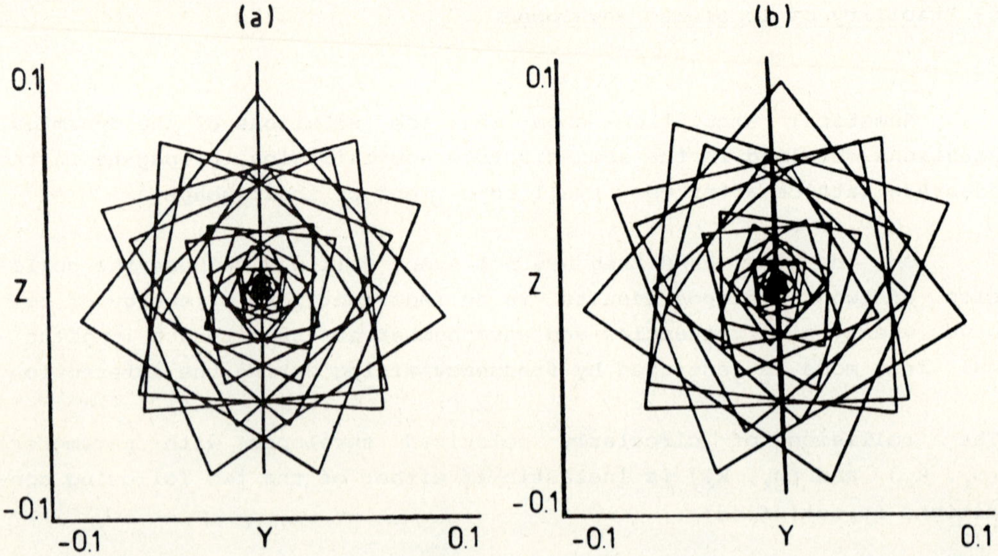

Figure 5: Plot of the transverse displacement (a) before and (b) after a collision a circularly and a linearly polarized envelopes showing the variation of polarization.

6- Conclusion.

While circularly polarized envelopes are not qualitatively modified by the inclusion of the longitudinal degree of freedom, linearly polarized waves that can be considered as bound states of two circularly polarized waves, are much affected. The stability properties are influenced by the existence of a resonance that produces linear longitudinal radiations when two circularly polarized modes interact.

References.

[1] F. FALK, Ginzburg-Landau theory of static domain walls in shape memory alloys, Z. Phys. B -Condensed Matter 51 (1983), 177-185.

[2] F. FALK, Ginzburg-Landau theory and solitary waves in shape memory alloys, Z. Phys. B -Condensed Matter 54 (1984), 159-167.

[3] O.B. GORBACHEVA and L.A. OSTROVSKY, Non linear vector waves in a mechanical model of a molecular chain, Physica D8 (1983), 223-228.

[4] S. CADET, Transverse envelope solitons in an atomic chain, Phys. Lett. A 121 (1987), 77-82.

[5] , Propagation and interactions of non linear shear waves in a discrete lattice, to appear in Wave Motion (1988).

[6] , Coupled transverse-longitudinal envelope modes in an atomic chain, J. Phys. C -Solid State Physics 20 (1987) L803-L811.

[7] T. KAWAHARA, The derivative expansion method and non linear dispersive waves, J. Phys. Soc. Jpn. 35 (1973) 1537-1544.

[8] N. FLYTZANIS, St. PNEVMATIKOS and M. REMOISSENET, Kink, breather and asymmetric envelope or dark solitons in non linear chains: I Monoatomic chain, J. Phys. C - Solid State Physics 18 (1985) 4603-4629.

[9] J. SATSUMA and N. YAJIMA, Initial value problems of one dimensional self modulation of non linear waves in dispersive media, Progr. Theor. Phys. Suppl. 55 (1974), 284-306.

[10] A.C. SCOTT, F.Y.F. CHU and D.W. McLAUGHLIN, The soliton: a new concept in applied science, Proceedings of the IEEE 61 (1973), 1443-1483.

[11] M. OIKAWA and N. YAJIMA, A perturbation approach to non linear systems II Interaction of non linear modulated waves, J. Phys. Soc. Jpn. 37 (1974), 486-496.

THE COMPUTATION OF THE AUSTENITIC-MARTENSITIC PHASE TRANSITION

CHARLES COLLINS AND MITCHELL LUSKIN
School of Mathematics and Minnesota Supercomputer Institute
Minneapolis, Minnesota 55455.

1. Introduction. The austenitic-martensitic phase transition is a solid-solid phase transition usually characterized by a decrease in crystal symmetry at a transition temperature. The indium-thallium alloy with composition near 20 atomic % thallium exhibits such a phase transition from a high temperature solid phase with cubic symmetry (austenite) to a low temperature solid phase with tetragonal symmetry (martensite) [5]. It is observed that the martensitic phase is often "finely twinned" along planes related to the crystal lattice and that an austenite/finely twinned martensite interface is observed at the transition temperature [5]. In this paper, we give results which demonstrate that the experimentally observed phenomena of fine scale martensitic twinning and the austenite/finely twinned martensite interface can be obtained by numerical computations based on the Ericksen theory for cubic-tetragonal phase transitions [8,9,10].

The Ericksen theory gives a free energy density for first-order cubic-tetragonal phase transitions. The moduli have been determined by R. D. James to match the linear elastic moduli for the austenitic and the martensitic phase at the transition temperature.

The theory developed by Ball and James [2] and Chipot and Kinderlehrer [6,15] explains the interesting phenomena that occur because the bulk energy functional is not lower semi-continuous. In particular, minimizing sequences give fine scale twinning and an austenite/finely twinned martensite interface.

We have developed numerical algorithms and a three-dimensional finite element code to compute approximate minima for the bulk energy functional. Our numerical results give twinning on the scale of the grid. Since the bulk energy functional is not lower semi-continuous, this is compatible with the expectation that minimizing sequences will have oscillatory deformation gradients.

Alt, Hoffman, Niezgódka, and Sprekels have given a numerical study of a one-dimensional mathematical model for the dynamics of the austenitic-martensitic phase transition [1]. Silling has used a non-physical bulk energy in two space dimensions to numerically simulate martensitic transformation and twinning [16].

2. Continuum Theory. Let the reference configuration for the crystal, $\Omega \subseteq \mathbf{R}^3$, be undistorted austenite at the critical temperature, θ_c. Let $y(x) : \Omega \to \mathbf{R}^3$ be the

This research was supported by the National Science Foundation, Grants DMS 835-1080 and DMS 871-8881 , the Cray Research Foundation, ARO Grant DAAL03-88-K0170, and the Minnesota Supercomputer Institute.

deformation. As usual, we assume that $y(x)$ is continuous, injective, and orientation preserving [7]. The deformation gradient is given by $F(x) = \nabla y(x)$ and the right Cauchy-Green strain tensor is given by $C = F^T F$ [13]. We denote the temperature by θ and the free energy per unit volume by $\phi(F, \theta)$. The bulk energy for the deformation $y(x)$ at the temperature θ is then given by

$$\mathfrak{J}(y) = \int_\Omega \phi(\nabla y(x), \theta) dx.$$

We want our constitutive equation to be invariant under observer changes [13], so we want ϕ to satisfy

$$\phi(F, \theta) = \phi((F^T F)^{1/2}, \theta) \tag{2.1}$$

and we define

$$W(C, \theta) = \phi(C^{1/2}, \theta) = \phi(F, \theta).$$

Since the reference configuration has cubic symmetry, we want W to satisfy

$$W(R_i C R_i^T, \theta) = W(C, \theta) \tag{2.2}$$

for $R_i \in \mathcal{G}$ where $\mathcal{G} = \{R_1, \ldots, R_{24}\}$ is the cubic symmetry group of proper rotations.

In an unstressed state, the Cauchy-Green strain for the martensitic phase is given by one of the variants

$$\begin{aligned} C_1 &= C_1(\theta) = \operatorname{diag}(1 + 2\epsilon, 1 - \epsilon, 1 - \epsilon), \\ C_2 &= C_2(\theta) = \operatorname{diag}(1 - \epsilon, 1 + 2\epsilon, 1 - \epsilon), \\ C_3 &= C_3(\theta) = \operatorname{diag}(1 - \epsilon, 1 - \epsilon, 1 + 2\epsilon), \end{aligned}$$

where $\epsilon = \epsilon(\theta)$. Note that

$$\{R_i C_1 R_i^T | i = 1, \ldots, 24\} = \{C_1, C_2, C_3\}.$$

We also define the right stretch tensors

$$U_i = C_i^{1/2}.$$

Our energy density must predict that in an unstressed state the austenitic phase is stable for $\theta > \theta_c$, the austenitic phase and the martensitic phase can coexist for θ near θ_c, and the martensitic phase is stable for $\theta < \theta_c$. More precisely, we must have that

$$W(C, \theta) > W(I, \theta) \quad \text{for } C \neq I,\ \theta > \theta_c, \tag{2.3}$$

$$W(C, \theta) > W(C_1, \theta) = W(C_2, \theta) = W(C_3, \theta) \quad \text{for } C \neq C_1, C_2, C_3,\ \theta < \theta_c, \tag{2.4}$$

and

$$W(C,\theta_c) > W(I,\theta_c) = W(C_1,\theta_c) = W(C_2,\theta_c) = W(C_3,\theta_c) \qquad (2.5)$$
$$\text{for } C \neq C_1, C_2, C_3, I.$$

Ericksen has proposed the following energy density for a constrained elastic crystal [10]

$$\begin{aligned} W(C,\theta) &= b(\theta)J + c(\theta)K + d(\theta)J^2 \\ J &= \frac{1}{6}\{(\lambda_1-1)^2 + (\lambda_2-1)^2 + (\lambda_3-1)^2\} \\ K &= \frac{1}{2}(\lambda_1-1)(\lambda_2-1)(\lambda_3-1), \\ \lambda_1 &= C_{11}, \quad \lambda_2 = C_{22}, \quad \lambda_3 = C_{33} \end{aligned}$$

subject to the constraints

$$C_{12} = C_{23} = C_{13} = 0, \quad \text{tr } C = 3.$$

Ericksen has shown that the coefficients b, c, and d can be chosen so that W satisfies (2.1)-(2.5). Ericksen also proposed to the authors that the constraints could be replaced by adding penalty terms to the energy density to obtain

$$\begin{aligned} W(C,\theta) = {} & bJ + cK + dJ^2 + \\ & \frac{e}{2}(C_{12}^2 + C_{13}^2 + C_{23}^2 + C_{21}^2 + C_{31}^2 + C_{32}^2) + f(\text{tr } C - 3)^2, \end{aligned}$$

$$\begin{aligned} J &= \frac{1}{6}\{(\lambda_1-1)^2 + (\lambda_2-1)^2 + (\lambda_3-1)^2\}, \\ K &= \frac{1}{2}(\lambda_1-1)(\lambda_2-1)(\lambda_3-1), \end{aligned} \qquad (2.6)$$

where the λ_i are modified so that

$$\lambda_i = \frac{3C_{ii}}{\text{tr } C} \qquad \text{for } i = 1,2,3.$$

R. D. James has determined coefficients for W so that (2.1)-(2.5) are satisfied, so that W matches available experimental linear elastic moduli at $\theta = \theta_c$ for the martensitic phase ($F = U_1$), and so that C_1, the Cauchy-Green strain for the unstressed martensitic phase, matches experimental data at $\theta = \theta_c$ for the indium-thallium alloy with 20.5 atomic % thallium. James' coefficients also match experimental data [5] for the effect of uniaxial stress on transition temperature. James' moduli are (θ in $°C$ and moduli in gigapascals)

$$\begin{aligned} b &= 0.38 + (1.22 \times 10^{-3})(\theta - 70) \\ c &= -29.23 \\ d &= 562.13 \\ e &= 3.26 \\ f &= 5.25 \end{aligned}$$

The critical temperature, the temperature at which (2.5) holds, is $\theta_c = 70$ for these coefficients.

3. Internally Twinned Martensite. For an unstressed solid at $\theta < \theta_c$ the above theory allows the existence of minimum energy deformations which have deformation gradients which are discontinuous across twin planes [2,6,8,9,10,15]. Ball and James [2] have shown that there exists a proper rotation, R, such that

$$RU_1 = U_3 + a \otimes n$$

where

$$\begin{aligned} n &= \frac{1}{\sqrt{2}}(e_1 + e_3), \\ a &= \frac{3\sqrt{2}\epsilon}{2+\epsilon}((1-\epsilon)^{1/2}e_1 - (1+2\epsilon)^{1/2}e_3). \end{aligned} \tag{3.1}$$

Planes orthogonal to n are twin planes. It then follows that

$$y(x) = U_3 x + a \int_0^{x \cdot n} \beta(t)dt$$

where $\beta(t)$ is a characteristic step function, is a deformation which has minimum energy and which has a discontinuous deformation gradient.

It is also easy to see that our bulk energy functional cannot be lower semi-continuous. For instance, if $\beta(t)$ is periodic of period 1, then the sequence of minimizers

$$k^{-1}y(kx) \rightharpoonup (U_3 + \lambda\, a \otimes n)x$$

weak-* in $W^{1,\infty}(\Omega)$ as $k \to \infty$ where

$$\lambda = \int_0^1 \beta(t)dt,$$

but $(U_3 + \lambda\, a \otimes n)x$ is not a deformation of minimum energy.

Now for $\theta < \theta_c$ and $0 < \lambda < 1$ the bulk energy function, $\mathfrak{J}(y)$, cannot attain its minimum value on the set of admissible deformations [3]

$$\begin{aligned} \mathcal{A} = \{ & y \in W^{1,\infty}(\Omega, \mathbf{R}^3) \,|\, \det(\nabla y) > 0 \ \text{ a.e. in } \Omega \\ & \text{and } y(x) = (U_3 + \lambda\, a \otimes n)x \text{ for } x \in \partial\Omega \}. \end{aligned}$$

This is because the deformation gradients of minimizing sequences attempt to take values which minimize the energy density. This results in fine twinning to give compatibility with the boundary conditions. The information to be obtained from the minimizing sequences for this problem can be summarized using the concept of the Young measure [6,15]. In fact, any minimizing sequence for this problem gives a unique, nontrivial Young measure which converges to a single laminate [3]. We are interested in the possibility of numerically computing minima of the bulk energy on finite-dimensional approximations of the set of admissible deformations as a means toward understanding the material microstructure given by minimizing sequences.

For our numerical experiments we let our reference configuration be

$$\Omega = \{x = (x_1, x_2, x_3) | 0 < x_i < 1 \text{ for } i = 1, 2, 3\}.$$

To construct our finite element approximation to $\mathcal{A}$ we let N be a positive integer, $h = 1/N$, and

$$\Omega_{ijk} = \{x = (x_1, x_2, x_3) | ih < x_1 < (i+1)h,\ jh < x_2 < (j+1)h,\ kh < x_3 < (k+1)h\}$$

for $i, j, k = 0, \ldots, N-1$. We then define the space of trilinear polynomial functions

$$Q_1 = \{\xi(x) | \xi(x) = \sum_{i,j,k=0}^{1} \alpha_{ijk} x_1^i x_2^j x_3^k \text{ for } \alpha_{ijk} \in \mathbf{R}^3\}$$

and the space of continuous, piecewise trilinear deformations by

$$\mathcal{M}_h = \{y(x) | y(x) \in C(\Omega, \mathbf{R}^3) \text{ and } y|_{\Omega_{ijk}} \in Q_1 \text{ for } i, j, k = 0, \ldots, N-1\}.$$

We approximate the set of admissible deformations by the finite-dimensional space

$$\mathcal{A}_h = \{y \in \mathcal{M}_h | y(x) = (U_3 + \lambda\, a \otimes n)x \text{ for } x \in \partial\Omega\}. \tag{3.2}$$

We then wish to compute $\tilde{y} \in \mathcal{A}_h$ such that

$$\mathcal{J}(\tilde{y}) \leq \mathcal{J}(y), \qquad \forall y \in \mathcal{A}_h. \tag{3.3}$$

Note that we have not required deformations in $\mathcal{A}_h$ to be orientation preserving ($\det \nabla y > 0$). This is acceptable since our computed solutions to (3.3) have all been orientation preserving. We have approximated the integrals in (3.3) by mid-point quadrature, and we have used the gradient method and the Fletcher-Reeves version of the conjugate gradient method [11,12] to compute minima of (3.3). Of course, both the gradient method and the conjugate gradient method can converge to local minima.

We were unable to obtain a fine scale twinned minima for (3.2). The reason seems to be that the twin planes (planes across which the deformation gradient is discontinuous) should be orthogonal to $n = \frac{1}{\sqrt{2}}(e_1 + e_3)$, but the deformations in $\mathcal{A}_h$ are not allowed to have discontinuities in their deformation gradients across planes orthogonal to n. However, we have obtained martensitic twinning on the scale of the mesh if the reference configuration is rotated so that the expected twin planes lie along planes for which the deformations in $\mathcal{A}_h$ can have discontinuous deformation gradients. This is easily done by rotating the reference configuration by $\frac{\pi}{4}$ about the x_2-axis. We shall continue to denote the coordinates in the new reference configuration by $x = (x_1, x_2, x_3)$ and the displacement by $y(x)$. We assume that the body in the new coordinates is described by

$$\Omega = \{x = (x_1, x_2, x_3) | 0 < x_i < 1 \text{ for } i = 1, 2, 3\}.$$

The energy density is now given by

$$\hat{\phi}(F, \theta) = \phi(FS, \theta)$$

where S is the rotation of $\frac{\pi}{4}$ about the x_2-axis, and the bulk energy is given by

$$\hat{\mathfrak{J}}(y) = \int_\Omega \hat{\phi}(\nabla y, \theta) dx.$$

The martensitic strains are now given by

$$\hat{C}_i = S C_i S^T$$

and

$$\hat{U}_i = S U_i S^T,$$

and we set

$$\hat{a} = Sa \text{ and } \hat{n} = Sn = e_1.$$

The set of admissible deformations for the continuous problem is now

$$\begin{aligned} \mathcal{A} = &\{y \in W^{1,\infty}(\Omega, \mathbf{R}^3) | \det(\nabla y) > 0 \text{ a.e. in } \Omega \\ &\text{and } y(x) = (\hat{U}_3 + \lambda \hat{a} \otimes \hat{n})x \quad \text{for } x \in \partial\Omega\} \end{aligned}$$

and the finite element subspace is

$$\mathcal{A}_h = \{y \in \mathcal{M}_h | y(x) = (\hat{U}_3 + \lambda \hat{a} \otimes \hat{n})x \quad \text{for } x \in \partial\Omega\}.$$

The problem is then to compute $\tilde{y} \in \mathcal{A}_h$ such that

$$\tilde{\mathfrak{J}}(\tilde{y}) \leq \tilde{\mathfrak{J}}(y), \qquad \forall y \in \mathcal{A}_h.$$

In order to see the effect of twinning more easily in our graphical output of the deformed state, we have replaced b by $16b$ and c by $4c$. This has the effect of replacing $\epsilon(\theta_c)$ by $4\epsilon(\theta_c)$. Thus, $\epsilon(\theta_c)$ for the new coefficients equals .104.

The most dramatic technique for showing the finely twinned structure we obtained from our numerical computation for this problem is to mark each cell in our reference configuration to denote whether the right Cauchy-Green strain at the mid-point is close to the austenitic state, $C = I$, or any of the martensitic variants, $C_i(\theta)$. In Figure 2 and

Figure 3 we show our results for the case $\lambda = 1/2$ and $\theta = 69$ (recall that $\theta_c = 70$) after 700 iterations of the gradient method with $h = 1/16$. The initial state is $y(x) = (\hat{U}_3 + \lambda \hat{a} \otimes \hat{n})x$. We consider a cell to be in the austenitic state if at the midpoint,

$$\|C - I\| \leq \text{minimum}\{\|C - C_1\|, \|C - C_2\|, \|C - C_3\|, (.02)^{1/2}\}$$

where the matrix norm is defined by

$$\|B\| = (\text{tr } B^T B)^{1/2} = \left(\sum_{i,j=1}^{3} B_{ij}^2 \right)^{1/2} .$$

We consider a cell to be in variant i of the martensitic state if at the midpoint

$$\|C - C_i\| \leq \text{ minimum}\{\|C - C_j\| \text{ for } j \neq i, \ \|C - I\|, \ (.02)^{1/2}\},$$

and we consider a cell to be not in a austenitic or a martensitic state if

$$\text{minimum}\{\|C - I\|, \|C - C_1\|, \|C - C_2\|, \|C - C_3\|\} \geq .02^{1/2}.$$

Figure 1 gives a key for all of the later figures.

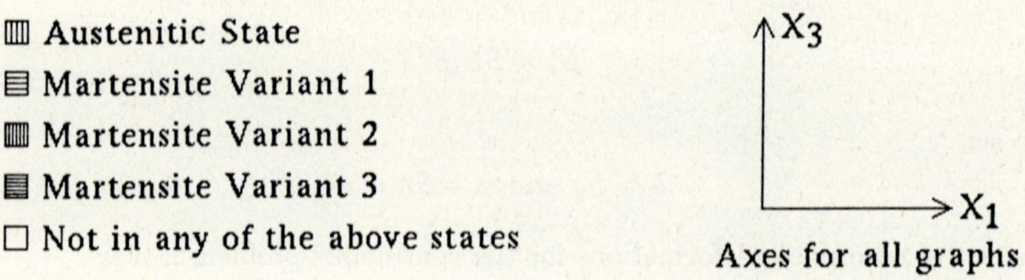

Figure 1. Key to later figures.

It is also informative to study the deformation of the crystal. In Figure 4 and Figure 5 we show the x_1-x_3 deformation of the crystal for the planar cross-sections $x_2 = \frac{3}{16}$ and $x_2 = \frac{1}{2}$. It is easy to see the shearing in the planar layers. It is also evident from Figures 2-5 that there is a planar defect in the crystal which is characterized by a change in the martensitic variant in the layers. This defect indicates that we are stuck in a local minimum. The final distribution of states is given by

austenitic state	221
martensite variant 1	1894
martensite variant 2	0
martensite variant 3	1889
not in any of the above states	92

We note that

$$\lambda_{\text{calc}} \equiv \frac{\# \text{ of states in martensite variant 1}}{\# \text{ of states in martensite}} = .50066$$

which is in excellent agreement with $\lambda = 1/2$.

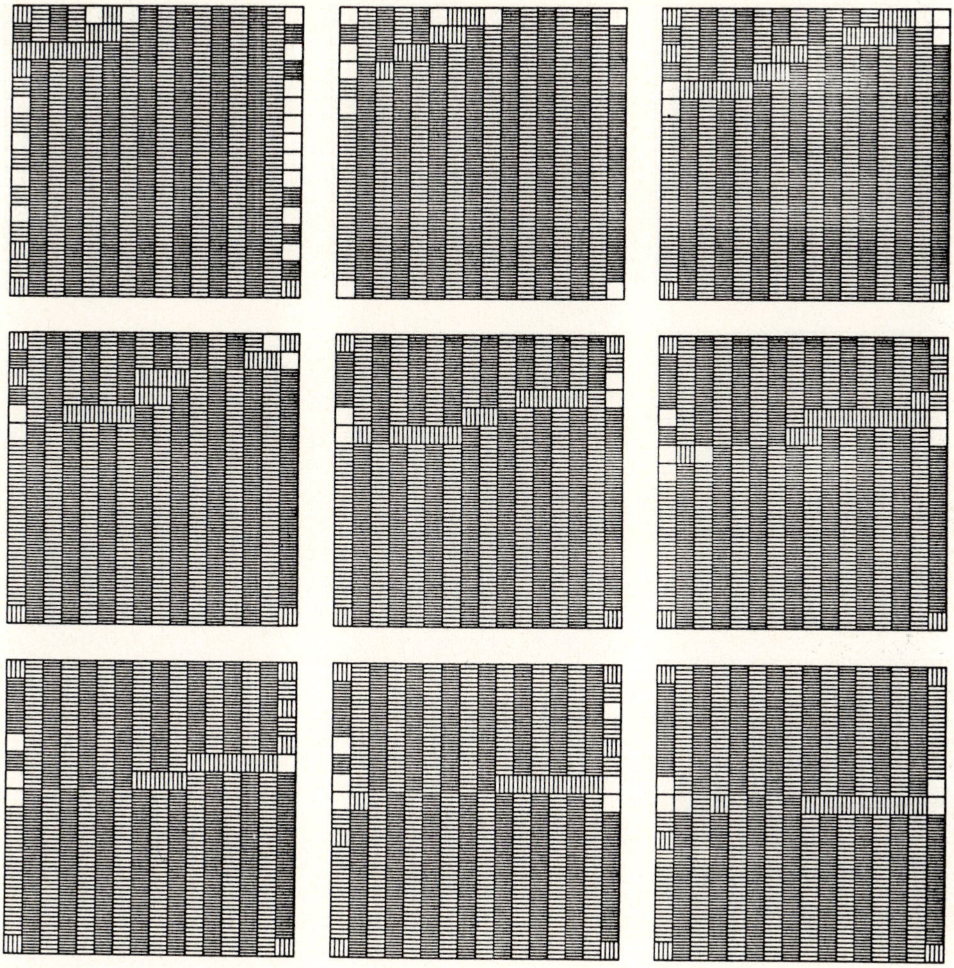

Figure 2. Planar cross-sections for $x_2 = (i + \frac{1}{2})h$ for $i = 0, \ldots, 8$ with $\lambda = 1/2$.

§4. **The Austenite/Finely Twinned Martensite Interface.** Ball and James have shown how the existence of an austenite/finely twinned martensite interface can be explained by the theory of finite elasticity and minimizing sequences [2]. To construct the interface, they show that there exists λ^* such that $0 < \lambda^* < 1$ and that there exists a proper rotation, Q, such that

$$\hat{U}_3 + \lambda^* \hat{a} \otimes \hat{n} = Q(I + \hat{b} \otimes \hat{m}) \tag{4.1}$$

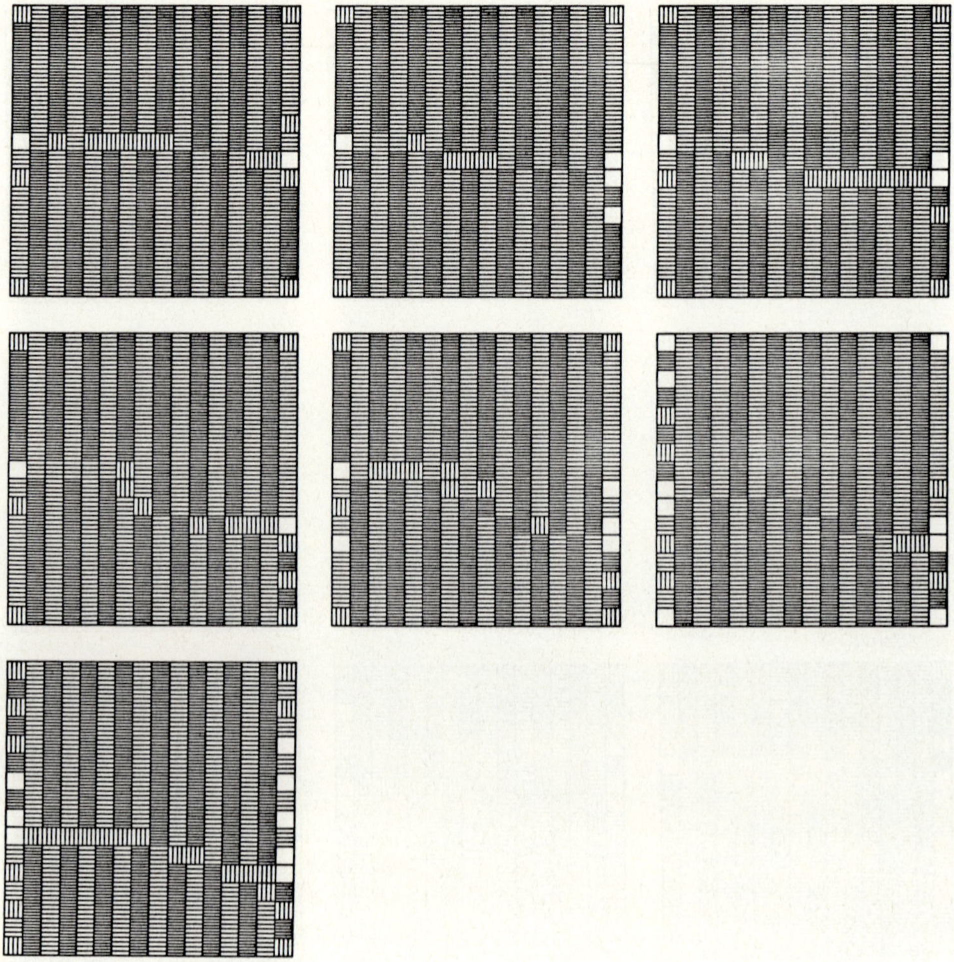

Figure 3. Planar cross-sections for $x_2 = (i + \frac{1}{2})h$ for $i = 9, \ldots, 15$ with $\lambda = 1/2$.

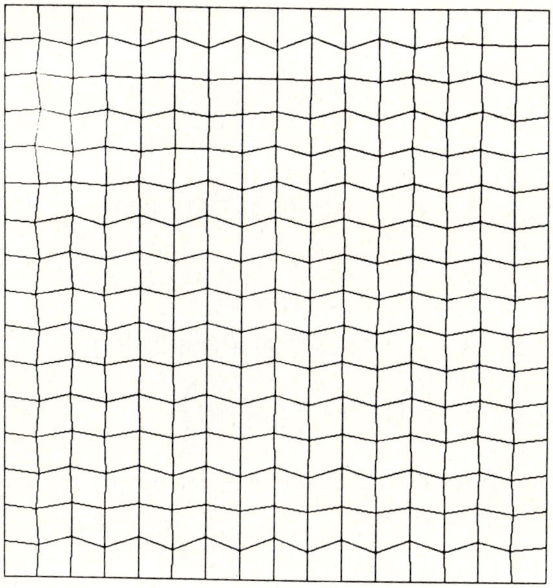

Figure 4. Planar cross-section of the deformation for $x_2 = \frac{3}{16}$ with $\lambda = 1/2$.

Figure 5. Planar cross-section of the deformation for $x_2 = 1/2$ with $\lambda = 1/2$.

where

$$
\begin{aligned}
\hat{m} &= Sm, \\
m &= \left(-\frac{1}{2}(1-\tau),\ 1,\ \frac{1}{2}(1+\tau)\right), \\
\hat{b} &= Sb, \\
b &= \left(-\frac{1}{2}\zeta(1-\tau),\ \beta,\ \frac{1}{2}\zeta(1+\tau)\right), \\
\tau &= (1-4\epsilon)^{1/2}, \\
\zeta &= \epsilon(1+(1+2\epsilon)^{1/2})^{-1}, \\
\beta &= -\,\epsilon(1+2\epsilon)^{1/2}(1+(1+2\epsilon)^{1/2})^{-1}, \\
\lambda^* &= \frac{1}{2}\left[1-\left(1-\frac{4}{9}(2+\epsilon)\right)^{1/2}\right].
\end{aligned}
$$

As a consequence of (4.1) they show that at $\theta = \theta_c$, for any $\alpha \in \mathbf{R}$, there exists a minimizing sequence $y_k \in W^{1,\infty}(\Omega, \mathbf{R}^3)$ such that

$$\hat{\mathcal{J}}(y_k) \to 0$$

and

$$y_k \rightharpoonup y^* \qquad \text{weak-* in } W^{1,\infty}(\Omega, \mathbf{R}^3)$$

where

$$y^*(x) = \begin{cases} Qx & \text{for } x \cdot \frac{\hat{m}}{|\hat{m}|} < \alpha \\ \alpha Q \frac{\hat{m}}{|\hat{m}|} + (\hat{U}_3 + \lambda^* \hat{a} \otimes \hat{n})(x - \alpha \frac{\hat{m}}{|\hat{m}|}) & \text{for } x \cdot \frac{\hat{m}}{|\hat{m}|} > \alpha. \end{cases}$$

Further, outside of a boundary layer about $x \cdot \frac{\hat{m}}{|\hat{m}|} = \alpha$ whose width, $w(k)$, converges to zero as $k \to \infty$,

$$\nabla y_k = \begin{cases} Q & \text{for } x \cdot \frac{\hat{m}}{|\hat{m}|} < \alpha \\ \hat{U}_3 + \beta(k\, x \cdot \hat{n})\hat{a} & \text{for } x \cdot \frac{\hat{m}}{|\hat{m}|} > \alpha + w(k) \end{cases}$$

where β is a characteristic step function which takes only the values 0 and 1, which is periodic with period 1 and which satisfies

$$\lambda^* = \int_0^1 \beta(t)dt.$$

Thus, the crystal is in the austenite state for $x \cdot \frac{\hat{m}}{|\hat{m}|} < \alpha$ and is in a finely twinned martensite state for $x \cdot \frac{\hat{m}}{|\hat{m}|} > \alpha$. The plane of the interface satisfies $x \cdot \frac{\hat{m}}{|\hat{m}|} = \alpha$.

We now consider the minimization of $\hat{\mathcal{J}}(y)$ on the set of admissible deformations

$$\mathcal{A} = \{y \in W^{1,\infty}(\Omega, \mathbf{R}^3) \mid \det(\nabla y) > 0 \text{ a.e. in } \Omega,\ y(x) = y^*(x) \text{ for } x \in \partial\Omega\}.$$

We think that the minimum of $\hat{\mathcal{J}}$ is not attained in the set $\mathcal{A}$ [4], although this has not yet been rigorously proven. However, minimizing sequences give an austenite/finely twinned martensite interface as described above.

We have been able to numerically compute such an austenite/finely twinned martensite interface. We approximate $\mathcal{A}$ by the finite dimensional space.

$$\mathcal{A}_h = \{y \in \mathcal{M}_h | y(x) = y^*(x) \text{ for } x \text{ at interpolation points in } \partial\Omega\}.$$

As before, we then compute $\hat{y} \in \mathcal{A}_h$ such that

$$\hat{\mathcal{J}}(\hat{y}) \leq \hat{\mathcal{J}}(y), \qquad \forall y \in \mathcal{A}. \tag{4.2}$$

We have solved (4.1) by the gradient method with the initialization $y(x) = y^*(x)$ at interpolation points (grid points) in Ω.

In Figure 6 and Figure 7 we give our results for the right Cauchy-Green strain for $\alpha = 1/\sqrt{2}$, $h = 1/16$ and $\theta = \theta_c = 70$ after 350 iterations. The interface is clearly given as the boundary between an austenitic region and a finely twinned region. We note the presence of some twin planes oriented orthogonal to the twin planes which are orthogonal to $\hat{n}$. The final gradient distribution is given by

austenitic state	1492
martensite variant 1	992
martensite variant 2	0
martensite variant 3	1566
not in any of the above states	46

We note that

$$\lambda_{\text{calc}} \equiv \frac{\#\text{ of states in martensite variant 1}}{\#\text{ of states in martensite}} = .388$$

whereas $\lambda^* = .372$. In Figure 8 and 9 we give the deformation of the plane $x_2 = \frac{5}{16}$ and $x_2 = \frac{1}{2}$ for the above problem.

§5. Numerical Methods. In this section, we shall describe the gradient method and the conjugate gradient methods that we have used to find minima of the functional, $\hat{\mathcal{J}}$.

The gradient method starts with an initial guess, $y^\circ \in \mathcal{A}_h$, and determines an iteration sequence, $y^n \in \mathcal{A}_h$, such that (under appropriate conditions) $y^n \to \hat{y}$. After $y^n \in \mathcal{A}_h$ has been computed, we compute $y^{n+1} \in \mathcal{A}_h$ as follows. First, we compute the gradient, $g^n \in \mathcal{A}_h$, by

$$\int_\Omega g^n \cdot z\, dx = \int_\Omega \frac{\partial \hat{\phi}}{\partial F}(\nabla y^n, \theta) \cdot \nabla z\, dx, \qquad \forall z \in \mathcal{A}_h. \tag{5.1}$$

Then we compute $\rho_n \in \mathbf{R}$, $\rho_n > 0$ such that

$$\hat{\mathcal{J}}(y^n - \rho_n g^n) \leq \hat{\mathcal{J}}(y^n - \rho g^n) \qquad \forall \rho > 0. \tag{5.2}$$

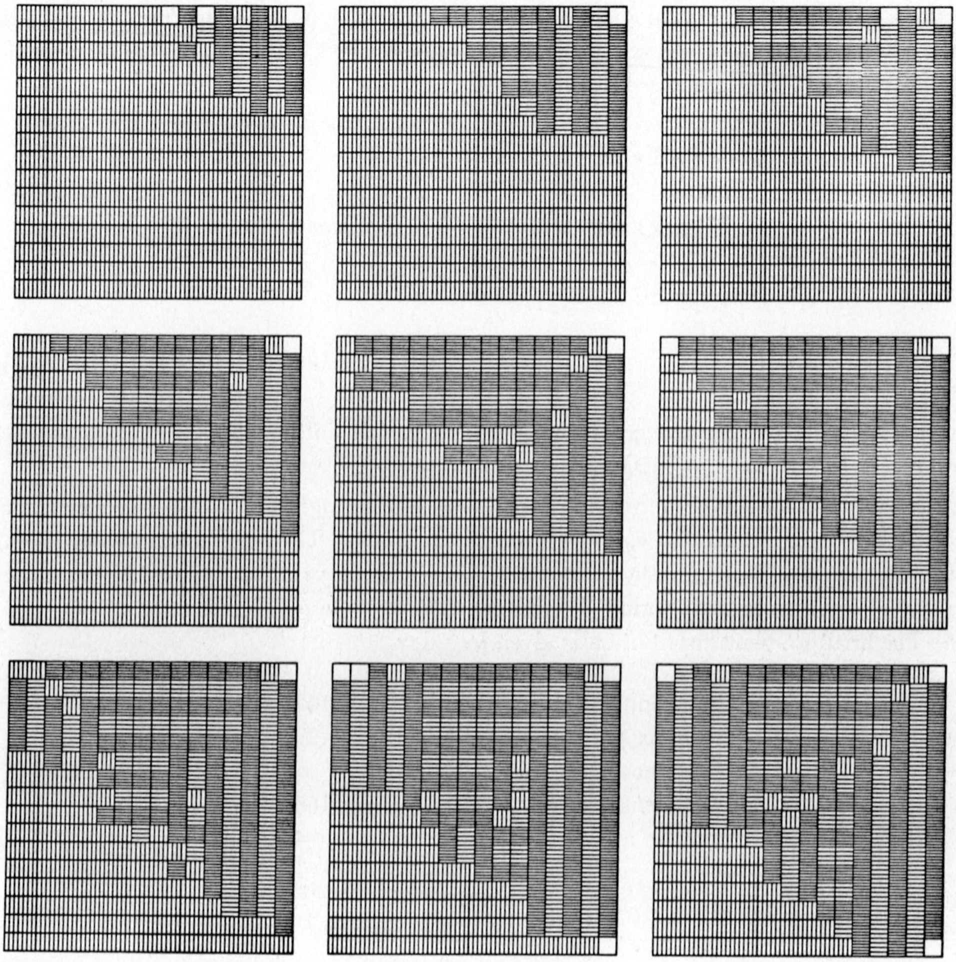

Figure 6. Planar cross-sections for $x_2 = (i+\frac{1}{2})h$ for $i = 0, \ldots, 8$ for the austenite/finely twinned martensite interface.

Finally, we set

$$y^{n+1} = y^n - \rho_n g^n.$$

We note that (5.2) is a univariate optimization problem for $\rho > 0$. In practice, we compute the first local minima for $\rho > 0$ in (5.2).

It is well-known that even for quadratic, convex functionals the gradient method can converge slowly [11]. For such problems the conjugate gradient method often converges in an order of magnitude fewer iterations. The Fletcher-Reeves variant of the conjugate gradient method is initialized by an initial guess $y^\circ \in \mathcal{A}_h$ and an initial search direction $d^\circ \equiv g^\circ \in \mathcal{A}_h$ is computed. After $y^n \in \mathcal{A}_h$ and the search direction $d^n \in \mathcal{A}_h$ have been

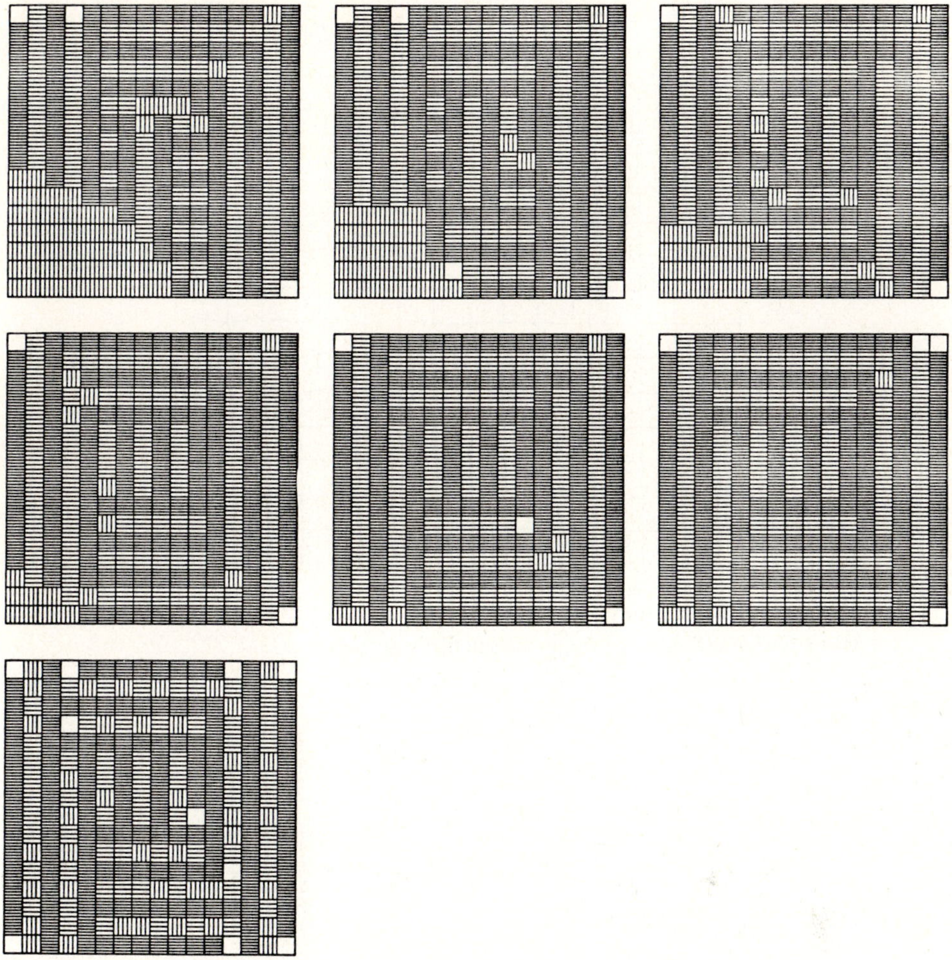

Figure 7. Planar cross-sections for $x_2 = (i + \frac{1}{2})h$ for $i = 9, \ldots, 15$ for the austenite/finely twinned martensite interface.

computed, we compute

$$y^{n+1} = y^n - \rho_n d^n$$

where $\rho_n \in \mathbf{R}$, $\rho_n > 0$ is the solution to the one-dimensional minimization problem

$$\hat{\mathcal{J}}(y^n - \rho_n d^n) \leq \hat{\mathcal{J}}(y^n - \rho d^n), \qquad \forall \rho > 0.$$

We then compute the new gradient, $g^{n+1} \in \mathcal{A}_h$, by

$$\int_\Omega g^{n+1} \cdot z \, dx = \int_\Omega \frac{\partial \hat{\phi}}{\partial F}(\nabla y^{n+1}, \theta) \cdot \nabla z \, dx, \qquad \forall z \in \mathcal{A}_h.$$

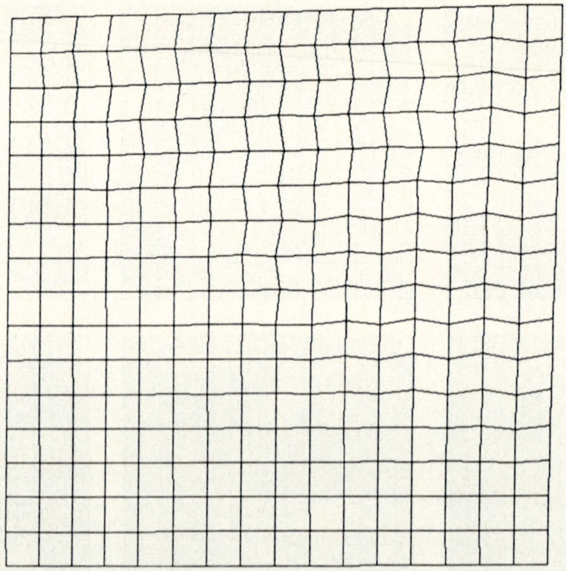

Figure 8. Planar cross-sections of the deformation for $x_2 = 5/16$ for the austenite/finely twinned martensite interface.

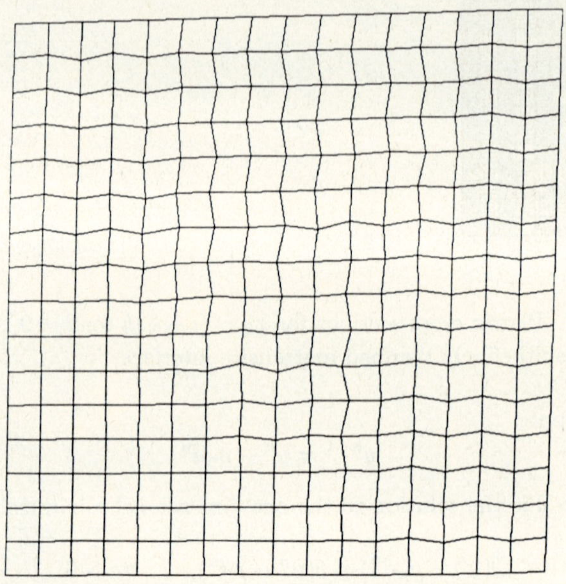

Figure 9. Planar cross-sections of the deformation for $x_2 = 1/2$ for the austenite/finely twinned martensite interface.

The new search direction, $d^{n+1} \in \mathcal{A}_h$, is then computed by

$$d^{n+1} = g^{n+1} + \lambda_n d^n$$

where

$$\lambda_n = \left[\int_\Omega g^{n+1} \cdot g^{n+1} dx\right] / \left[\int_\Omega g^n \cdot g^n \, dx\right]$$

Our computational experiments have show that the energy of the iterates for the conjugate gradient method initially decays faster than the energy of the iterates for the gradient method. However, the gradient method reaches our criterion for convergence as fast as the conjugate gradient method. In Figure 10, we give a comparison of the decay of the energy for iterates of the gradient method and the conjugate gradient method for the internally twinned martensite problem with $\lambda = 1/2$ as described in section 3.

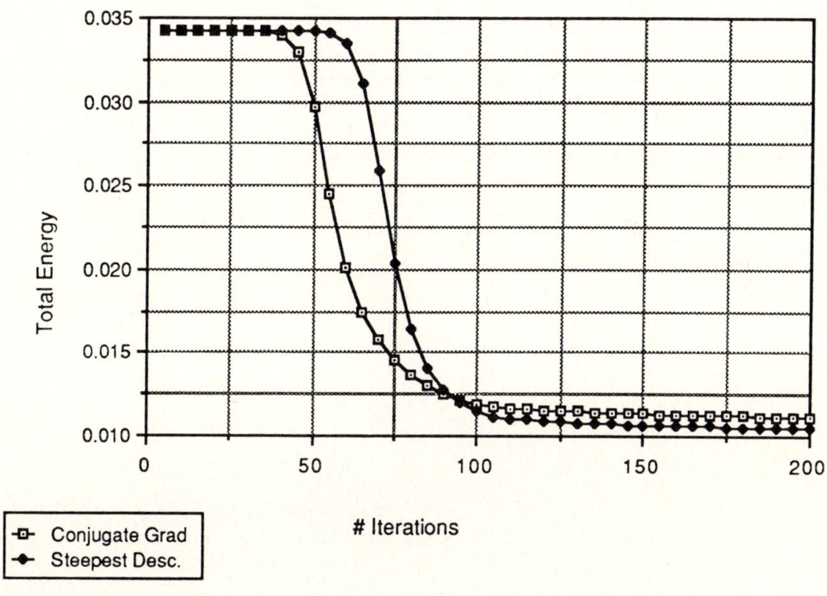

Figure 10. Comparison of the decay of the energy for iterates computed with the gradient method and the conjugate gradient method for the internally twinned martensite problem with $\lambda = 1/2$.

REFERENCES

[1] F. W. ALT, K.-H. HOFFMANN, M. NIEZGÓDKA, J. SPREKELS, *A numerical study of structural phase transitions in shape memory alloys*, Preprint #90, Institut für Mathematik, Universitat Augsburg, 1985.

[2] J. M. BALL AND R. D. JAMES, *Fine phase mixtures as minimizers of energy*, Arch. Rational Mech. Anal., 100 (1987), pp. 13-52.

[3] J. M. BALL AND R. D. JAMES, *Experimental tests of a theory of fine microstruture*, preprint, August, 1988.

[4] J. M. BALL, R. D. JAMES, AND D. KINDERLEHRER, in these proceedings.

[5] M. W. BURKART AND T. A. READ, *Diffusionless phase change in the indium-thallium system*, Trans. AIME J. Metals, 197 (1953), pp. 1516-1524.

[6] M. CHIPOT AND D. KINDERLEHRER, *Equilibrium configurations of crystals*, to appear, Arch. Rational Mech. Anal.

[7] P. G. CIARLET, *Mathematical Elasticity, Volume 1: Three-Dimensional Elasticity*, North-Holland, Amsterdam, 1988.

[8] J. L. ERICKSEN, *Some phase transitions in crystals*, Arch. Rational Mech. Anal., 73 (1980), pp. 99-124.

[9] J. L. ERICKSEN, *Some constrained elastic crystals*, (ed. J. M. Ball), Oxford University Press, 1987, pp. 119-137.

[10] J. L. ERICKSEN, *Constitutive theory for some constrained elastic crystals*, Int. J. Solids and Structures, 22 (1986), pp. 951-964.

[11] P. GILL, W. MURRAY, AND M. WRIGHT, *Practical Optimization*, Academic Press, London, 1981.

[12] R. GLOWINSKI, *Numerical Methods for Nonlinear Variational Problems*, Springer-Verlag, New York, 1984.

[13] M. E. GURTIN, *Topics in Finite Elasticity*, SIAM, Philadelphia, 1981.

[14] R. D. JAMES, *Microstructure and weak convergence*, in Material Instabilities in Continuum Mechanics and Related Problems, (ed. J. M. Ball), Oxford University Press, 1987, pp. 175-196.

[15] D. KINDERLEHRER, *Remarks about equilibrium configurations of crystals*, in Material Instabilities in Continuum Mechanics and Related Problems (ed. J. M. Ball), Oxford University Press, 1987, pp. 217-242.

[16] S. A. SILLING, *Phase changes induced by deformation in isothermal elastic crystals*, preprint.

Theory of Diffusionless Phase Transitions

Richard James

Department of Aerospace Engineering and Mechanics
University of Minnesota
Minneapolis, Minnesota

David Kinderlehrer

School of Mathematics
University of Minnesota
Minneapolis, Minnesota

1 Introduction

The aim of this report is to explore a constitutive theory able to predict macroscopic state functions and the detailed microstructure of a crystal which suffers a diffusionless solid-solid transformation involving a change in symmetry. A principal feature of this theory is that we are led to consider energy functionals which fail to be lower semi-continuous, with respect to an appropriate notion of weak convergence. In these circumstances, the infimum of energy may be achieved only in some generalized sense while a minimizing sequence may develop successively finer oscillations reminiscent of a finely twinned microstructure.[1]

To understand very briefly the consequences of material symmetry in this regard, let us consider an energy density $W(F)$ defined for an elastic body in three dimensional space at some fixed temperature. The condition of frame indifference is that

$$W(QF) = W(F) \qquad \text{for} \qquad Q^TQ = \mathbb{1},\ \det Q = 1,\ \text{and}\ \det F > 0. \qquad (1.1)$$

The variant structure of certain crystals, for example, many ionic solids, shape memory alloys, and ferroelectrics, suggests that coexistent symmetry related phases may be present in an configuration. An example of this is twinning. A deformation A is *symmetry related* to F provided

[1] This research was supported under NSF MSM 86 12420 and by the NSF and ASOFR under DMS 87 1881

$$A = QFH \tag{1.2}$$

for some proper rotation Q and some H in a given "symmetry group" and

$$W(A) = W(F) . \tag{1.3}$$

The two variants coexist in a configuration when there is a piecewise linear deformation $y(x)$ with, say,

$$\nabla y(x) = \begin{cases} F & \text{if } x \cdot n < 0 \\ A & \text{if } x \cdot n > 0 \end{cases} , \tag{1.4}$$

for some unit vector $n \in \mathbb{R}^3$. This is only possible if rank $(A - F) = 1$, provided $A \neq F$, as Hadamard pointed out, and in fact,

$$A = F + \alpha \otimes n \qquad \text{for some nonzero } \alpha \in \mathbb{R}^3.$$

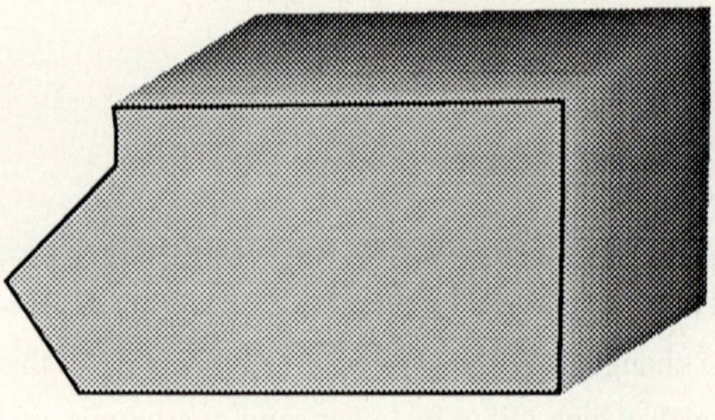

Fig. 1 Schematic drawing of a single crystal

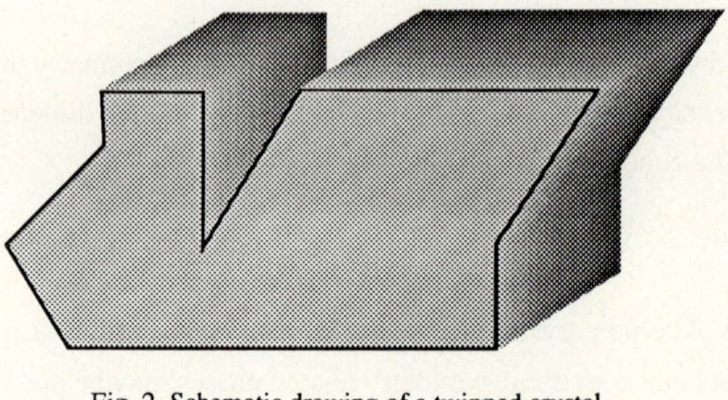

Fig. 2 Schematic drawing of a twinned crystal

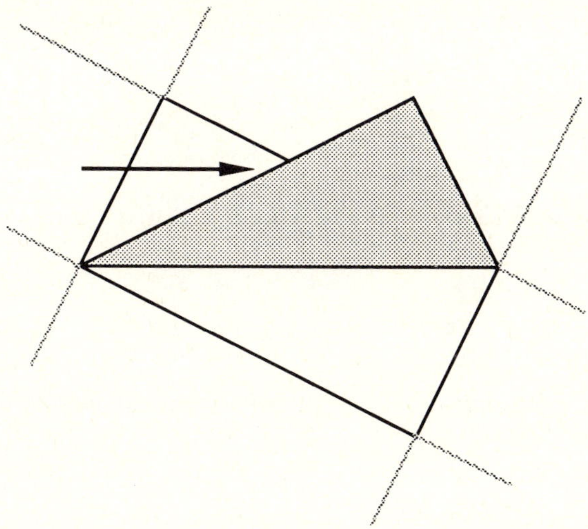

Fig 3. Illustration of the motion of a lattice element on the twin plane according to (1.4). The upper triangle undergoes simple shear which renders its deformed state congruent, in three dimensions, to its undeformed state.

Suppose, for clarity, that $W(F) = \min W = 0$. Then the function $f(t) = W(F + t\alpha \otimes n)$ vanishes together with its derivative at $t = 0$ and $t = 1$ and thus at some $\lambda \in (0,1)$,

$$f''(\lambda) = \frac{\partial^2 W}{\partial A_{ij} \partial A_{hk}} \alpha^i \alpha^h n^j n^k < 0 ,$$

unless f is constant.

Thus the familiar Legendre-Hadamard condition is violated for W. Of course, we know that lower semi-continuity of the functional

$$E(v) = \int_\Omega W(\nabla v)\, dx$$

implies the Legendre-Hadamard condition, cf. Morrey [51]. So our ability to understand variant structures in crystalline solids depends on our willingness to study functionals which are not lower-semicontinuous.

On the other hand, let us write

$$A = F(\mathbb{1} + a \otimes n) . \tag{1.5}$$

Given $\lambda \in (0,1)$, introduce the characteristic function $\chi(t)$ of the interval $(0,\lambda) \subset (0,1)$, extended periodically to $(-\infty,\infty)$, so $\chi(t) = \chi(t + k)$, $k \in \mathbb{Z}$. Let

$$u^k(x) \;=\; F\Big(x + \frac{1}{k}\int_0^{kx\cdot n} \chi(t)\,dt\;\; a\Big) \tag{1.6}$$

so that

$$F^k(x) \;=\; \nabla u^k(x) \;=\; F(1 \;+\; \chi(kx\cdot n)\, a \otimes n) \tag{1.7}$$

$$= \begin{cases} F & \text{if}\;\; \chi \;=\; 0 \\ A & \text{if}\;\; \chi \;=\; 1 \end{cases} \;.$$

Now the (F^k) do not converge in a usual pointwise sense, but they do converge in the weak* topology. Indeed,

$$F^k \;\rightarrow\; (1 - \lambda)F \;+\; \lambda A \qquad \text{in}\;\; L^\infty(\Omega)\;\; \text{weak*}, \tag{1.8}$$

when $\Omega \subset \mathbb{R}^3$ is bounded, since the sequence of periodic functions $\chi(kt)$, $0 \le t \le 1$, converges weak* to its mean value λ. Inspection of the (F^k) shows that it represents a sequence of twinned states with successively finer structure. From the viewpoint of energy,

$$W(F^k(x)) \;=\; \begin{cases} W(F) & \text{if}\;\; \chi \;=\; 0 \\ W(A) & \text{if}\;\; \chi \;=\; 1 \end{cases}$$

$$= 0, \quad \text{for all k.} \tag{1.9}$$

We see in this way that the macroscopic limit deformation, which itself is not a minimizer of energy, is obtained by successive shearing of minimum energy configurations. In the sequel we shall illustrate how the energy of the limit configuration may be interpreted as $\lim_{k\to\infty} W(F^k) = 0$ by use of the parametrized measure or Young measure [74], cf. also [67]. This measure serves as an accounting device to record the oscillations of the process through which the limit configuration is achieved.

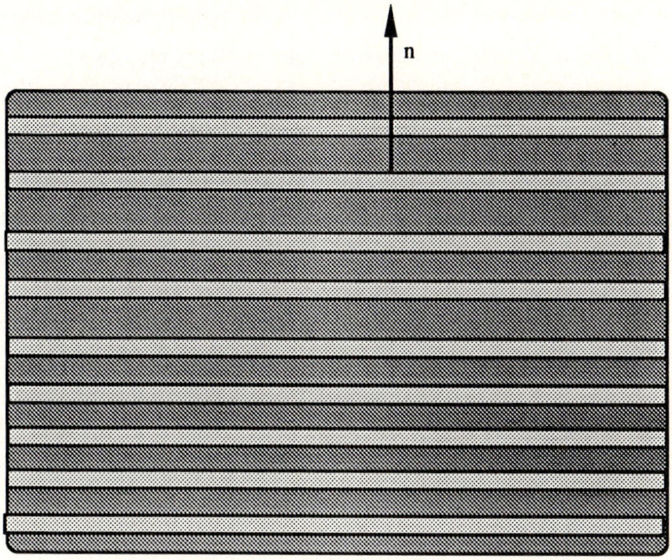

Fig. 4 Illustration of fine phase behavior, as given by F^k in (1.9).

The phenomenological theory we are about to describe has its origins in the work of Ericksen [21-29]. Our idea of equilibrium follows from Gibbs [34]. Important contributions are due to Gurtin [35], Müller [52], Müller and Wilmanski [53], Parry [54], and Pitteri [61-63]. Some of our thoughts about phase transitions have been motivated by studying [10,12,16,20,71,72,73]. This report is based primarily on [6],[14]. The paper by Ball [5] has some connections with this one. Computational work related to the theory and analysis presented here has been undertaken by Collins and Luskin [17]. Other work of possible interest to the reader includes [38-41,42,43].

2 Constitutive equations

We now describe how particular energy density functions $W(F,\theta)$ are obtained. Sometimes we shall depress the dependence of the energy on temperature θ. We begin with lattice considerations. Our idea of a crystal is that it consists of a regular lattice consisting of identical atoms, molecules, or more general symmetry elements $L(e_1,e_2,e_3)$ determined by linearly independent lattice vectors $\{ e_i \}$ according to the rule

$$L(e_1,e_2,e_3) = \left\{ x \in \mathbb{R}^3 : x = \mu^i e_i , \ \mu^i \in \mathbb{Z} \right\}. \tag{2.1}$$

The summation convention is understood to be in force. A common procedure involves assigning a central force potential and then summing the contributions of all the energies due to pairs of atoms in a large volume. The temperature dependence of the energy density is then calculated by recognizing that the atoms vibrate about the definite positions in $L(e_1,e_2,e_3)$. Standard statistical arguments then provide a temperature correction. It is well known that this procedure fails in that it predicts the Cauchy relations which are not satisfied by many materials of interest in studies of phase transformations. As discussed by Christian [16, Chapter 5] modern attempts to improve the calculation have been largely unsuccessful, and, in fact, many of these calculations once again give back the Cauchy relations!

Recognizing that for numerous calculations of the type illustrated in the introduction we will not need the detailed form of W, we adopt a more modest but general point of view. A derivation of W from first principles will have the property that once the positions of the lattice elements are known, the energy density is determined. Following this line of thought, we postulate the existence of $\Phi(e_1,e_2,e_3,\theta)$ which represents the energy per unit volume of a lattice $L(e_1,e_2,e_3)$ at temperature θ. The danger in this procedure is that two different sets of lattice vectors may generate the same lattice, i.e., the same atomic positions. This possibility is addressed by a classical theorem in crystallography, cf. [24] or [29], stating that the energy does not depend on the choice of lattice basis, leading to the invariance statement

$$\Phi(\mu_1^j e_j, \mu_2^j e_j, \mu_3^j e_j,\theta) = \Phi(e_1,e_2,e_3,\theta) \qquad \text{for all } M = (\mu_i^j) \in \mathbb{Z}^9,\ \det M = \pm 1. \tag{2.2}$$

We may pass to a continuum theory by adopting the Cauchy - Born Rule [24,27]. For this, we fix a basis of reference lattice vectors $\{ e_i^o \}$ and define

$$W(F,\theta) = \frac{(Fe_1^o \wedge Fe_2^o)\cdot Fe_3^o}{(e_1^o \wedge e_2^o)\cdot e_3^o}\,\Phi(Fe_1,Fe_2,Fe_3,\theta), \qquad \det F > 0. \tag{2.3}$$

The factor in front of Φ merely converts the energy per unit volume to an energy per unit reference volume to be more consistent with standard continuum mechanics. This factor is really just $\det F$. With this rule, the Ω introduced in §1 becomes identified with the lattice $L(e_1^o, e_2^o, e_3^o)$.

Clearly W should be frame indifferent at each fixed θ and so satisfy (1.1). As well, it inherits the invariance from (2.2)

$$W(FH,\theta) = W(F,\theta), \qquad \text{for all } H \in \mathbb{G}(\{ e_i^o \}), \text{ where} \tag{2.4}$$

$$\mathbb{G}(\{ e_i^o \}) \; = \; E^o \, GL(\mathbb{Z}^3) \, (E^o)^{-1} \; , \qquad (2.5)$$

where E^o is the matrix whose columns are $\{ e_i^o \}$. Note that if F_o minimizes W at some fixed θ, then so does $F_o\mathbb{G}(\{ e_i^o \})$. No matter what the choice of F_o, $\det F_o > 0$, the set $F_o\mathbb{G}(\{ e_i^o \})$ is infinite. So $W(A,\theta)$ has an infinite number of potential wells, and there are $H^k \in \mathbb{G}(\{ e_i^o \})$, $k = 1,2,3,...$ such that $| H^k F_o | \to \infty$. In fact, if $\mathbb{1} + a \otimes n \in \mathbb{G}(\{ e_i^o \})$, then $W(A(\mathbb{1} + \lambda a \otimes n))$ is a periodic function of λ. If we associate F with the actual deformation of the body, it is clear that we will not be able to deform the body into one of these far away potential wells without compromising our notion of the lattice. That is, we expect dislocations to appear and to move with such large deformations, and they should surely contribute to the energy density.

Calculations of energy minima for an energy displaying the invariance (2.4) and (2.5) by Chipot and Kinderlehrer [14] and Fonseca [31,32] support the inference above. However, they are useful for special loading devices, as we point out in the next section, and give some general insight into the nature of instabilities. At the present time, energy densities satisfying (2.4),(2.5) are the only ones for which the *relaxed energy* have been calculated.

Since dislocation movement appears unimportant for many transformation, it is plausible that the remarks above may be irrelevant. What is working remarkably well for many transformations is the assumption that W has the form (2.4) but that it has a domain D restricted to include only a finite number of potential wells. In this framework, we choose $\{ e_i^o \}$ to be associated with the undistorted more symmetric phase and assume that $\mathbb{1}$ belongs to D. The objective is to delimit an appropriate neighborhood D of $\mathbb{1}$. A remarkably useful guide for this purpose has been nonlinear elasticity theory in its more classical form with invariance governed by the subgroup $\mathbb{G}(\{ e_i^o \}) \cap SO(3)$, which is essentially a point group, in (2.4). A distillation of many recent calculations is this: choose D to include the transformed lattice, i.e., to include a matrix U such that $L(Ue_1^o, Ue_2^o, Ue_3^o)$ is the transformed lattice, require that D be invariant under $\mathbb{G}(\{ e_i^o \}) \cap SO(3)$ in the sense that $F \in D \Leftrightarrow F \in D\,[\mathbb{G}(\{ e_i^o \}) \cap SO(3)]$, and finally require that D not include any potential well minima not forced to be in it by the above criterion. Details of derivations of this type are given by Ball and James [7], Bhattacharya [11], Ericksen [30].

To understand what emerges of these considerations, let us consider a simple cubic to tetragonal transformation.

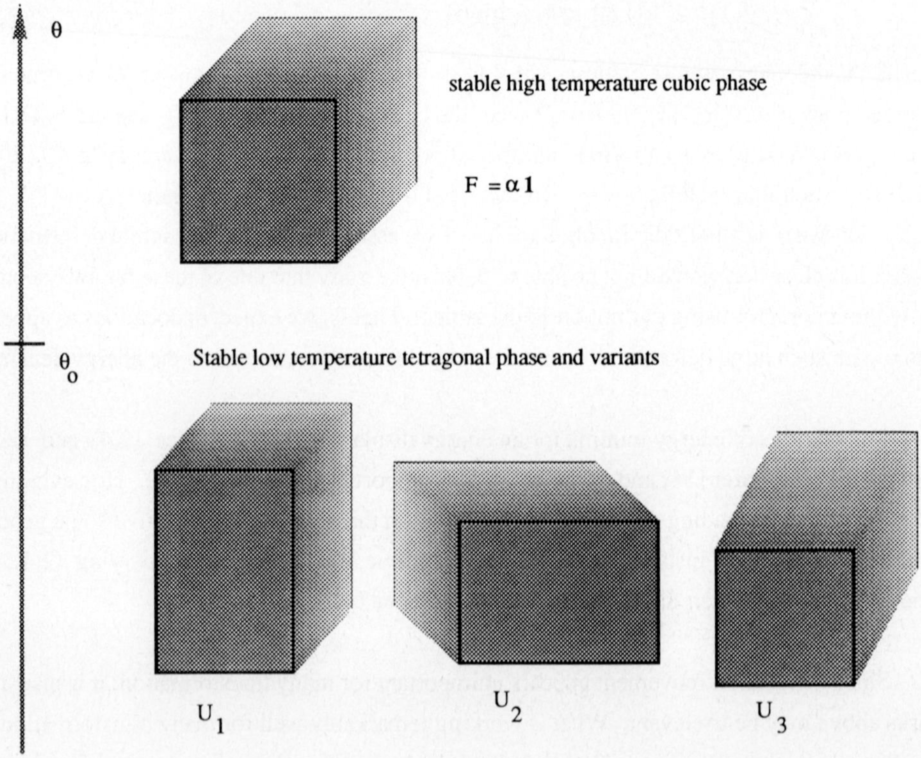

Fig. 5 Illustration of a cubic/tetragonal transition

In this situation,

for $\theta > \theta_o$: $W(A,\theta)$ has strict minima on the orbit $\alpha(\theta)SO(3)$ and
for $\theta < \theta_o$: $W(A,\theta)$ has strict minima on the union of orbits

$$SO(3)\,U_1(\theta) \cup SO(3)\,U_2(\theta) \cup SO(3)\,U_3(\theta),$$

with $U_i(\theta) = \eta_1(\theta)\mathbb{1} + (\eta_2(\theta) - \eta_1(\theta))\, e_i^o \otimes e_i^o$ (no sum), where the $\{ e_i^o \}$ are orthonormal and the scalar valued functions α, η_1, and η_2 are positive and continuous. Since W is continuous, we necessarily have at the critical temperature

$$W(\mathbb{1},\theta_o) = W(U_1,\theta_o) = W(U_2,\theta_o) = W(U_3,\theta_o) \qquad (2.6)$$

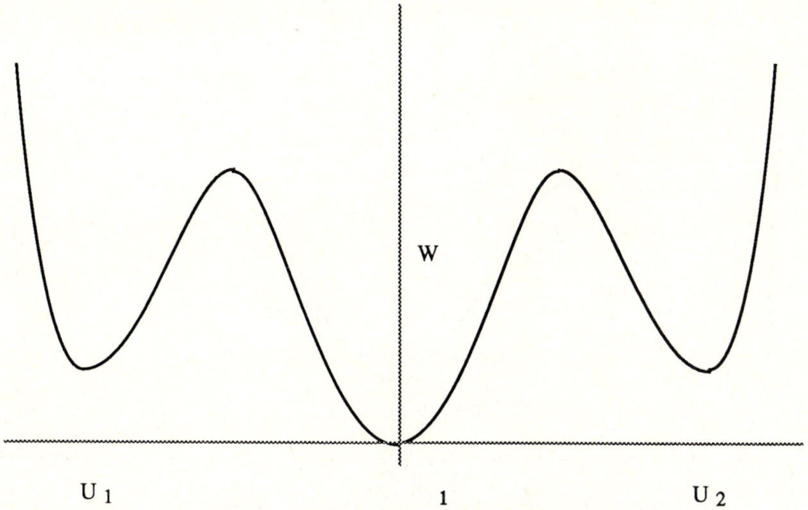

Fig. 6 Cubic phase at minimum energy above transition temperature.

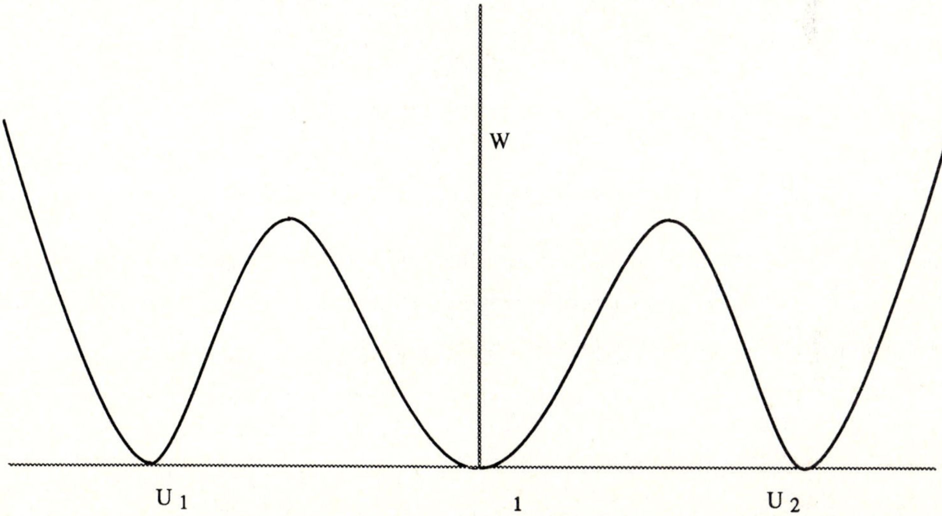

Fig. 7 Cubic and tetragonal phases at minimum energy at critical temperature

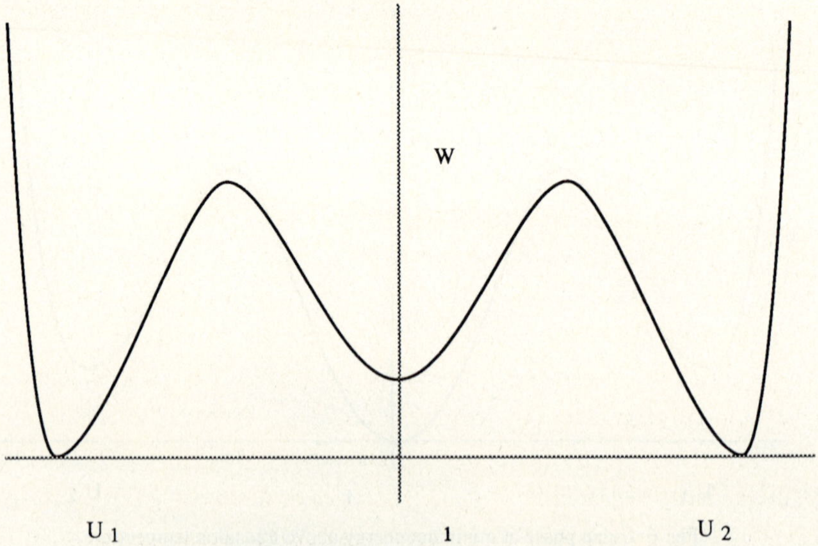

Fig. 8 Tetragonal phase at minimum energy below transition temperature

With these assumptions, as we shall describe in the ensuing sections, we may consider

$$\inf_{A} \int_{\Omega} W(\nabla y,\theta)\, dx \tag{2.7}$$

in a suitable class of functions A for various values of θ, and we are quickly led to consider the possibility that (2.7) has minimizers of the type (1.4), whatever that may mean. To describe these minimizers we frequently use the terminology

$\alpha(\theta)SO(3)$ *Austenite well,*

$SO(3)\, U_1(\theta) \cup SO(3)\, U_2(\theta) \cup SO(3)\, U_3(\theta)$ *Martensite wells.*

The typical situation for measured transformation strains is $\eta_1 \neq 1$ and $\eta_2 \neq 1$. In several known cases $(\eta_1)^2\eta_2 \approx 1$. We assume that $\eta_1 \neq 1 \neq \eta_2$. An easy calculation shows that there are no solutions of (1.4) with F and A on the same well. However we do find *rank - one connections* between any pair of martensite wells. The definite planes which can be planes of discontinuity, those of normal n in (1.4), arise from this calculation and provide an immediate first comparison between theory and experiment. In the cubic to tetragonal case, the planes are of the {110} family and agree with those observed below the transformation temperature in many cubic to tetragonal transformations.

However, there are no rank-one connections between austenite and martensite, whereas crystals at $\theta = \theta_0$ are observed to contain certain austenite/finely twinned martensite interfaces. These can be understood in terms of certain minimizing sequences similar to those constructed in §1. We describe them in §4.

A notable feature of this development is that it contains no contribution for the surface energy which might be assigned to such surfaces of discontinuity. Thus some features of the microstructure which appear fine in the microscope are modelled as infinitely fine by the present theory. Typical interlaminar distances that occur in austenite/finely twinned martensite interfaces range from about 10 μm down to a few atomic spacings, depending on the material. In a given material the fineness is approximately fixed. It is easy to ascribe such *limited fineness* to a small energy per unit area on the surfaces of gradient discontinuity, which would function to penalize the formation of arbitrarily fine laminates. Scaling calculations based on this idea do give reasonable qualitative results [6]. They predict, for example, that sufficiently small crystals will not contain an austenite/finely twinned martensite interface, an observed fact. A reasonable point of view is that in sufficiently large crystals, for example of diameter > 1 mm, the surface energy serves just to pick out some fine scale, while the minimization of the bulk energy determines the kinematical properties of the microstructure on that scale. Furthermore, it is believed that the minimizing sequences of the bulk energy give reasonable values of the bulk properties like deformation, free energy, and stress, cf. §4.

Of course, bulk theory alone will be useless for determining the scale of fineness or the precise details of the microstructure. It also will not give a complete description of growth phenomena. Recent work on this general problem is due to Gurtin [36]. Parry [55] and Fonseca [33] have discussed surface energy based on Herring's interpretation of the Wulff construction [37]. Recently, some thought has been given to the implications of minimizing the sum of bulk and surface energy [44].

Our lattice considerations here have been restricted to simple lattices where only mechanical interactions are presumed. Similar considerations are applicable to lattices with shifts or to polarizable dielectrics where W also depends on electrical polarization. In the latter case, the energy of a configuration as measured by (2.7) must include a potential which delivers the Maxwell stress. The theory for quartz is worked out in [40]. The paper [42] contains a discussion of ferroelectrics.

3 Principle of virtual work

Common prejudice is to represent an equilibrium configuration at fixed temperature by a deformation $y(x)$ of a reference domain Ω which renders stationary the energy functional; that is,

$$\delta \int_\Omega W(\nabla y,\theta)\, dx = 0 . \tag{3.1}$$

In the present situation, it is not easy to justify this since any affine $y(x)$ satisfies (3.1) and we have seen that many are unstable. We prefer to think in terms of stability criteria, but, unfortunately, we are lacking guidelines in this regard. About these issues, much has been written, [2,3,9,14,57,58]. To surmount these difficulties, at least from the analytical point of view, we shall resort to direct methods.

Provisionally, this issue may be separated into two parts. The first is to seek the minimum energy available to a configuration, which is the topic of this section. Second, we may ask about the deformation which delivers this minimum, the values of various state functions, and the determination of fine scale or microstructural features. Recall that the two parts are separate because the functional of (3.1) is not lower semicontinuous. Hence, as illustrated in the previous sections, the loading may give rise to a macroscopic deformation which only represents a local spatial average over potential wells, whose energy need not be a minimum.

Given a domain $\Omega \subset \mathbb{R}^3$ with suitably smooth boundary and $y_o \in H^{1,\infty}(\Omega)$ with $\det \nabla y_o > 0$, let

$$A = A_\Omega(y_o) = \{ u \in H^{1,\infty}(\Omega) : \det \nabla u > 0 \text{ and } u = y_o \text{ on } \partial\Omega \} \tag{3.2}$$

and set

$$E_\Omega(y_o,\theta) = \inf_A \int_\Omega W(\nabla u,\theta)\, dx . \tag{3.3}$$

Here we briefly discuss the evaluation of (3.3). An analysis undertaken in this spirit is sometimes called "relaxation," and one objective is to replace, at least temporarily, the original density W by a density $W^\#$ which is lower semicontinuous and for which

$$E_\Omega(y_o,\theta) = \inf_A \int_\Omega W^\#(\nabla u,\theta)\, dx . \tag{3.3}$$

Necessarily, if (3.3) holds, then in particular whenever $y_o(x) = F_o x$ is affine, $\det F_o > 0$, we have that

$$W^{\#}(F_o,\theta)\,|\Omega| = \inf_{H} \int_{\Omega} W(F_o + \nabla\zeta,\theta)\, dx \ , \tag{3.4}$$

$$H = \{ \zeta \in H_0^{1,\infty}(\Omega) : \det(F_o + \nabla\zeta) > 0 \} .$$

The proof of (3.4) follows from an easy adaptation of the argument of [6] . Were the kinematical constraint on the competing functions absent, it would be possible to show that (3.3) is verified for any data $y_o \in H^{1,\infty}(\Omega)$, but we are only able to establish this in limited, albeit useful, circumstances. On a deeper mechanical level, it is very difficult to actually know the function $W^{\#}$ in terms of W.

In his discussion of the Gibbs phenomenon for thermoelastic solids, Ericksen [25] observed that the energy of a homogeneously deformed body in equilibrium with a heat bath might tend to assume the value of a certain subenergy which he defined as

$$\varphi(\det F,\theta) = \inf_{\det A = \det F} W(A,\theta) . \tag{3.5}$$

Ericksen illustrated that this was consistent with certain types of thermodynamical thought where critical temperatures state functions were computed or measured in terms of specific volume alone, ignoring other features of the deformation.

We are able to give an additional interpretation to this subenergy for a crystalline solid when (2.4) and (2.5), namely, invariance under the full lattice group, is assumed. Here it is possible to show that

$$W^{\#}(F,\theta) = \varphi^{**}(\det F,\theta), \quad \det F > 0, \tag{3.6}$$

where $\varphi^{**}(t,\theta)$ is the convexification of the function $\varphi(t,\theta)$ as a function of the real variable t. For the proof of this we refer to [14,32]. In this case, $W^{\#}$ is polyconvex, namely, a convex function of A, adj A, and det A, but for more restricted invariance groups, we do not expect this.

Two examples where (3.3) holds may be worthwhile mentioning.

A. *Assume that* W *satisfies* (2.4) *and* (2.5) *so that* $W^\#$ *is given by* (3.6). *If* $y_o \in C^1(\bar{\Omega})$ *satisfies*

$$\det \nabla y_o \;=\; \upsilon \qquad in\ \Omega ,$$

where υ *is constant, then*

$$E_\Omega(y_o,\theta) \;=\; \inf_A \int_\Omega \varphi^{**}(\det \nabla u,\theta)\,dx \;=\; \varphi^{**}(\upsilon,\theta)\,|\,\Omega\,| .$$

B. *Assume that* $\inf W = 0$ *and that* $E_\Omega(y_o,\theta) = 0$. *Then*

$$E_\Omega(y_o,\theta) \;=\; \inf_A \int_\Omega W^\#(\nabla u,\theta)\,dx \;=\; 0 .$$

Although it appears a banality, a consequence of part B is that it assists us to calculate *all* possible deformation gradients F for which $W^\#(F,\theta) = 0$ when W is assumed to have a given well structure.

There is an extensive literature about relaxation and lower semicontinuity of variational integrals, [1,2,13,18,19,45-49,59,66].

4 Young measures: state functions, and microstructure

We now arrive at the place where we wish to relate the minimum energy of a configuration with the deformation which arises as the limit of a minimizing sequence. As we have suggested earlier, the technical device we adopt for this purpose is the Young measure, or parametrized measure. We shall give to it an interpretation in terms of local spatial averages and illustrate how it may be employed to analyze properties of equilibrium configurations. The use of Young measures in differential equations was first introduced by Tartar [68-70], especially in order to study hyperbolic conservation laws. It will serve us as an accounting tool to summarize the oscillatory properties of a minimizing sequence.

Let us assume that we are in the favorable situation where (3.3) holds, so the infimum of energy is the same as the infimum of the relaxed energy. Further, suppose we have in hand a sequence $y^k \in A_\Omega(y_o)$ such that

$$E_\Omega(y_o,\theta) \;=\; \lim_{k\to\infty} \int_\Omega W^{\#}(F^k,\theta)\,dx \;=\; \lim_{k\to\infty} \int_\Omega W(F^k,\theta)\,dx, \qquad (4.1)$$

$$F^k \;=\; \nabla y^k ,$$

and moreover

$$y^k \to y \qquad \text{in } H^{1,\infty}(\Omega) \text{ weak*}. \qquad (4.2)$$

By choice of a subsequence, if necessary, this will be the case provided $\| F^k \|_{L^\infty(\Omega)} \le C < \infty$ for some constant C.

First let us remind ourselves that (4.2) means

$$\left.\begin{aligned} &y^k \to y \qquad \text{uniformly in } \overline{\Omega} \text{ and} \\ &\int_E F^k dx \to \int_E F\,dx \qquad \text{for all (measurable) } E \subset \Omega, \end{aligned}\right. \qquad (4.3)$$

where $F = \nabla y$. Pointwise convergence of the sequence (F^k) will ordinarily fail.

A general fact is that given a sequence (f^k) which is bounded in $L^\infty(\Omega)$, it has a subsequence $(f^{k'})$ and an $f \in L^\infty(\Omega)$ such that

$$f^{k'} \to f \qquad \text{in } L^\infty(\Omega) \text{ weak*},$$

namely, $(4.3)_2$. So given a continuous function $\psi(A)$ defined on 3×3 matrices, the sequence $(\psi(F^k))$ is bounded in $L^\infty(\Omega)$ and thus for some subsequence, which we continue to denote by $(\psi(F^k))$, there is a $\overline{\psi} \in L^\infty(\Omega)$ for which

$$\psi(F^k) \to \overline{\psi} \qquad \text{in } L^\infty(\Omega) \text{ weak*}. \qquad (4.4)$$

Inasmuch as usually the function $\overline{\psi}$ is not the same as $\psi(F)$, what is it? This is the question to which the Young measure responds. In the present framework, there is a family $(\nu_x)_{x \in \Omega}$ of probabability measures defined on the set $\mathbb{M}$ of matrices, and depending measurably on $x \in \Omega$, such that

$$\overline{\psi}(x) = \int_M \psi(A)\, d\nu_x(A) \quad \text{a.e. in } \Omega. \tag{4.5}$$

Since $F^k \to F$ in $L^\infty(\Omega)$ weak*, we have immediately that

$$F(x) = \int_M A\, d\nu_x(A) \qquad \text{a.e. in } \Omega. \tag{4.6}$$

In the simple but important case of (1.6), one easily verifies that

$$\nu_x = (1-\lambda)\,\delta_F + \lambda\,\delta_A .$$

The oscillatory nature of the sequence (u^k) is recorded in the variant proportion λ. In fact, it is obvious that given $x_o \in \Omega$,

$$1-\lambda = \lim_{\rho\to\infty} \lim_{k\to\infty} \frac{1}{|B_\rho|} \left| \{ x \in B_\rho(x_o) \colon \nabla u^k(x) = F \} \right| .$$

The analysis of the Young measure $(\nu_x)_{x\in\Omega}$ rests upon understanding the role of weak* continuous and lower semicontinuous functions $\psi(A)$. In general, these are the minors of the matrix A, explicitly, the nineteen functions A, $\operatorname{adj} A = \det A\, A^{-T}$, and $\det A$. This means that

$$\operatorname{adj} F^k \to \operatorname{adj} F \quad \text{and} \quad \det F^k \to \det F \qquad \text{in } L^\infty(\Omega) \text{ weak*}.$$

Employing our notation (4.5), we may write

$$\left.\begin{aligned} \operatorname{adj} F(x) &= \int_M \operatorname{adj} A\, d\nu_x(A) \quad \text{and} \\ \det F(x) &= \int_M \det A\, d\nu_x(A) \end{aligned}\right\} \quad \text{a.e. in } \Omega. \tag{4.7}$$

Now $W^\#$ is lower semicontinuous, so we obtain from (4.1) and our hypothesis (3.3) that

$$\int_\Omega W^\#(F,\theta)\, dx \le \lim_{k\to\infty} \int_\Omega W^\#(F^k,\theta)\, dx$$

$$= E_\Omega(y_o,\theta) \;\le\; \int_\Omega W^\#(F,\theta)\,dx$$

so that

$$\int_\Omega W^\#(F,\theta)\,dx \;=\; \lim_{k\to\infty} \int_\Omega W^\#(F^k,\theta)\,dx \;=\; \lim_{k\to\infty} \int_\Omega W(F^k,\theta)\,dx\;. \qquad (4.8)$$

A consequence of this is that the sequences $(W^\#(F^k,\theta))$ and $(W(F^k,\theta))$ both converge weak* to $W^\#(F,\theta)$, so we discover a twentieth weak* continuous function, one particular to the minimizing sequence. Again employing our notation (4.5),

$$W^\#(F(x),\theta) \;=\; \int_{\mathbb{M}} W^\#(A,\theta)\,d\nu_x(A) \;=\; \int_{\mathbb{M}} W(A,\theta)\,d\nu_x(A) \qquad \text{a.e. in } \Omega \qquad (4.9)$$

Since $W^\# \le W$, we deduce that

$$\operatorname{supp} \nu_x \;\subset\; \{\, A \in \mathbb{M}:\; W^\#(A,\theta) \;=\; W(A,\theta) \,\}, \qquad (4.10)$$

which is our first statement about the Young measure associated to the minimizing sequence (y^k). It will turn out that the algebraic structure of the set on the right in (4.10) may then limit or determine possible configurations and their microstructure. For example, in the austenite/martensite transition described in §2, the Young measure $(\nu_x)_{x \in \Omega}$ corresponding to any energy minimizing configuration at critical temperature θ_o will have

$$\operatorname{supp} \nu_x \;\subset\; \text{austenite well} \,\cup\, \text{martensite wells}\,.$$

Special choices of ψ in (4.5) provide us with the state functions of the configuration. For example, set $S(A,\theta) = \partial W(A,\theta)/\partial A$, the Piola - Kirchhoff stress. We might write that

$$\overline{S}(x,\theta) \;=\; \int_{\mathbb{M}} S(A,\theta)\,d\nu_x(A) \qquad \text{in } \Omega\,. \qquad (4.11)$$

This has a variational justification if $\det F(x) \ge c > 0$ in Ω. In this case,

$$\frac{d}{d\varepsilon}\int_\Omega\int_{\mathbb{M}} W(A+\varepsilon\nabla\zeta)\, d\nu_x(A)\, dx\,\Big|_{\varepsilon=0} = \int_\Omega\int_{\mathbb{M}} S(A)\cdot\nabla\zeta\, d\nu_x(A)\, dx$$

$$\text{for } \zeta\in H_0^{1,\infty}(\Omega). \tag{4.12}$$

Then $W - W^\#$ achieves its minimium value on $\operatorname{supp}\nu_x$, so if $W^\#$ is differentiable, with $S^\#(A) = \partial W^\#(A,\theta)/\partial A$, it follows from (4.9) that

$$\overline{S}(x,\theta) = \overline{S^\#}(x,\theta) = S^\#(F(x)) \text{ a.e. in } \Omega, \tag{4.13}$$

and this expression may be substituted into (4.12). Thus the state of stress in the body may always be described by a "relaxed" field equation,

$$-\operatorname{div} S^\#(F(x)) = 0 \text{ a.e. in } \Omega.$$

Of course, the second derivatives of W with respect to A do not agree with the second derivatives of $W^\#$. One discussion of linearization in this context and its relationship to homogenization is given in [15].

A second example of a state function easily computable is the specific heat at constant volume,

$$C_\upsilon = -\frac{1}{\theta}\frac{\partial^2 W}{\partial\theta^2}.$$

Again in this case, since $\overline{W}(x,\theta) = W^\#(F(x),\theta)$,

$$C_\upsilon = -\frac{1}{\theta}\frac{\partial^2 W^\#(F(x),\theta)}{\partial\theta^2} = -\frac{1}{\theta}\int_{\mathbb{M}}\frac{\partial^2 W}{\partial\theta^2}(A,\theta)\, d\nu_x(A). \tag{4.14}$$

In the special case of the symmetry of (2.4) and (2.5), where $W^\#(F,\theta) = \varphi^{**}(\det F,\theta)$,

$$C_\upsilon = -\frac{1}{\theta}\frac{\partial^2\varphi^{**}(\det F(x),\theta)}{\partial\theta^2}, \tag{4.15}$$

which depends only on the specific volume. In particular, the points where the right hand side is not differentiable, which are usually the transition temperatures like θ_o in (2.6), are determined by the specific volume alone. This is consistent with some views of the matter and some experimental practice, cf. eg. Pippard [60].

The examples above may be termed "macroscopic" in the sense that the results do not depend on the particular Young measure $(\nu_x)_{x \in \Omega}$. Let us now briefly discuss some "microstructural" properties, which do depend on the particular Young measure. This will also introduce some curious analytical questions as well.

There are, in essence, two equivalent ways in which the Young measure may be considered a local spatial average, Ball [5,§4] and [14,Theorem 7.3]. Given $a \in \Omega$ and $\rho > 0$, define the measure $\nu^k_{a,\rho}$ by

$$\langle \nu^k_{a,\rho}, \psi \rangle = \frac{1}{|B_\rho|} \int_{B_\rho(a)} \psi(F^k(x))\, dx \,. \tag{4.16}$$

This defines a continous linear functional on $C(K)$ for $K = \{ A : |A| \leq C \}$, cf. (4.2) et. seq. This measure is a probability measure giving the distribution of values of $F^k(x)$ in $B_\rho(a)$. By the weak convergence of the sequence (F^k),

$$\nu^k_{a,\rho} \to \nu_{a,\rho} \qquad \text{weak* as measures as } k \to \infty, \text{ where}$$

$$\langle \nu_{a,\rho}, \psi \rangle = \frac{1}{|B_\rho|} \int_{B_\rho(a)} \int_{\mathbb{M}} \psi(A)\, d\nu_x(A)\, dx \,. \tag{4.17}$$

By Lebesgue's Theorem,

$$\langle \nu_{a,\rho}, \psi \rangle \to \langle \nu_a, \psi \rangle \qquad \text{as } \rho \to 0 \quad \text{a.e. in } \Omega,$$

from which it follows by the separability of $C(K)$ that

$$\nu_{a,\rho} \to \nu_a \qquad \text{weak* as measures, a.e. in } \Omega. \tag{4.18}$$

These obvious remarks have the consequence that we may restrict ν_a for almost every $a \in \Omega$. In other words, by choosing a diagonal sequence of (k,ρ), one verifies that the family of measures

$$(\mu_x)_{x \in \Omega}, \quad \text{where} \quad \mu_x = \nu_a \quad \text{for all } x \in \Omega, \tag{4.19}$$

is also a parametrized measure which is determined by a sequence of deformation gradients. Another view of this is given by Matos [50].

The sequence $(\nu_{a,\rho})$ admits a convergent subsequence at every $a \in \Omega$, even if this is not an obvious spatial average. This will be useful in what follows.

We turn now to a few specific examples of the analysis of Young measures. As our first example, we shall show that a Young measure $(\nu_x)_{x \in \Omega}$ supported on a single well $SO(3)F_o$ for each $x \in \Omega$ reduces to a *constant* Dirac mass.

We may assume that $F_o = \mathbb{1}$ for the argument that follows by a change of variables. Suppose that $F(x) = \nabla y(x)$ is the underlying deformation in the sense of (4.6) as usual. Then

$$F(x) \;=\; \int_{SO(3)} Q \, d\nu_x(Q) \qquad \text{and}$$

$$\operatorname{adj} F(x) \;=\; \int_{SO(3)} Q \, d\nu_x(Q) \;=\; F(x)\,,$$

since $\operatorname{adj} Q = Q$ when Q is a proper rotation. Finally,

$$\det F(x) \;=\; \int_{SO(3)} d\nu_x(Q) \;=\; 1.$$

Thus

$$F(x) \;=\; \det F(x)\, F(x)^{-T} \;=\; F(x)^{-T}\,, \tag{4.20}$$

so $F(x)$ is a rotation. Now $SO(3) \subset$ sphere of radius $\sqrt{3}$ in $\mathbb{R}^9$, which is strictly convex, so the only way (4.20) can hold is if

$$\nu_x \;=\; \delta_{F(x)} \quad \text{a.e. in } \Omega.$$

A more efficient and general proof of this is given in Ball [4].

The second part of the argument is due essentially to Reshetnyak [64,65]. Of course we know that any smooth deformation gradient $F(x)$ satisfying $F(x) \in SO(3)$ must be constant. Let

$$U(A) = \frac{1}{2} |A|^2 - \det A , \tag{4.21}$$

which has $U_A(A) = A - \det A$. Since $\det A$ is a null-lagrangian, the Euler equation of U is

$$-\operatorname{div} U_A(\nabla u) = \Delta u .$$

If Q is any rotation, $\det Q = 1$, then $U_A(Q) = 0$, so for the $F(x)$ of (4.20), $U_A(F(x)) = 0$; whence,

$$\Delta y = -\operatorname{div} U_A(\nabla y) = 0 \text{ in } \Omega,$$

in the sense of H^1, say. Hence y is smooth and by the remark above about smooth deformations, $\nabla y(x) = F(x) = R$, for some rotation R, and

$$\nu_x = \delta_R \quad \text{a.e. in } \Omega. \tag{4.22}$$

This conclusion agrees with our abundant observations that pure austenite or a pure variant of martensite cannot form a microstructure. It is a sort of Young measure regularity theorem: knowledge of the support of the measure at each point determines its global behavior.

5 Young Measures: the Austenite /Martensite Transition

As a second example, consider the most common microstructure observed in a thermoelastic martensitic alloy at the transformation temperature, the austenite/finely twinned martensite interface, shown schematically in Fig. 9. As explained in §2, there are no rank-one connections between the austenite well and any of the martensite wells, so we cannot understand this microstructure in terms of piecewise differentiable functions. Moreover, the observation that the martensite variants are always observed to be finely twinned suggests the study of minimizing sequences. Our analysis is in the style of [6].

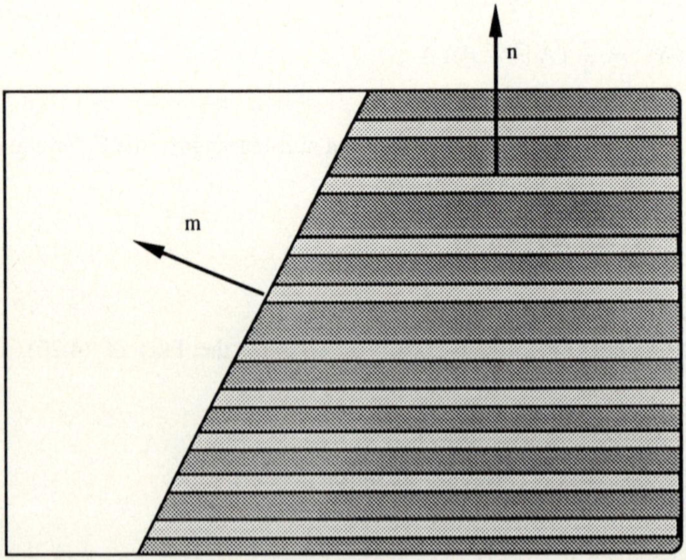

Fig. 9 Austenite/finely twinned martensite interface showing variants of martensite and the kinematical quantities m and n .

Observation of the photomicrographs quickly convinces one of a salient feature: there are essentially three deformation gradients involved, two of which belong to martensite wells and which differ by a rank-one matrix and one which belongs to the austenite well. Being noncomittal about the precise geometry of the microstructure, we therefore assume simply that we have in hand a sequence $y^k \to y$ in $H^{1,\infty}$ which is a minimizing sequence for the problem (2.7) and that its Young measure $(\nu_x)_{x \in \Omega}$, according to remark B of §3, satisfies

$$\text{supp}\, \nu_x \subset \{ M_1, M_2, C \} \quad \text{a.e. in } \Omega, \text{ where} \tag{5.1}$$

$$M_2 - M_1 = a \otimes n, \quad C - M_1 \neq \text{rank one}, \quad \text{and} \quad C - M_2 \neq \text{rank one}, \tag{5.2}$$

for vectors $a, n \in \mathbb{R}^3$. Of course we have in mind that $M_i = R_i U_i$, $i = 1,2$, and $C = R$, where R_1, R_2, and $R \in SO(3)$, and U_i are from (2.6).

Our interest lies in the observation that the presence of a microstructure involving the austenite and the martensite imposes restrictions on the matrices M_1, M_2, and C. We shall show that for some σ, $0 < \sigma < 1$,

$$\operatorname{rank}\,[\,C - ((1-\sigma)M_1 + \sigma M_2)\,] = 1, \tag{5.3}$$

that is, the two martensite variants are capable of averaging among themselves so that the result is kinematically compatible with the austenite. In addition, we illustrate how the presence of an austenite/martensite interface determines how the microstructure propagates in the sample. We shall show that there is a "propagation region", cf. (5.28), consisting of a mixture of austenite and finely twinned martensite which occurs with the fixed variant proportions $1-\sigma$ and σ. This is summarized by the reperesentation of the Young measure given in (5.29).

For simplicity, we shall assume that the reference configuration Ω is convex. Given the sequence (y^k), let $(\nu_x)_{x\in\Omega}$ denote the Young measure. By (5.1),

$$\nu_x = \lambda_1(x)\,\delta_{M_1} + \lambda_2(x)\,\delta_{M_2} + \lambda_3(x)\,\delta_C\,, \quad \sum \lambda_i(x) = 1 \quad \text{a.e. in } \Omega. \tag{5.4}$$

Step 1 If $\{\,0 < \lambda_3 < 1\,\}$ has positive measure, then (5.3) holds.

In this case, for almost every $a \in \{\,0 < \lambda_3 < 1\,\}$, we may consider the restricted Young measure ν_a. Let us set $\mu = \nu_a$. Since a may be chosen a Lebesgue point of λ_i, we may write

$$\mu = \tau_1\,\delta_{M_1} + \tau_2\,\delta_{M_2} + \tau_3\,\delta_C\,, \quad \sum \tau_i = 1, \quad 0 < \tau_3 < 1. \tag{5.5}$$

Let F denote the underlying deformation of μ. Again we have the relations for the minors of F, which are

$$\begin{aligned} F &= \tau_1 M_1 + \tau_2 M_2 + \tau_3 C\,, \\ \operatorname{adj} F &= \tau_1 \operatorname{adj} M_1 + \tau_2 \operatorname{adj} M_2 + \tau_3 \operatorname{adj} C\,, \quad \text{and} \\ \det F &= \tau_1 \det M_1 + \tau_2 \det M_2 + \tau_3 \det C\,. \end{aligned} \tag{5.6}$$

We now consider the identity

$$\operatorname{adj}\Big(\sum_1^3 \alpha_i A_i\Big) = \sum_1^3 \alpha_i \operatorname{adj} A_i - \frac{\alpha_1\alpha_2}{1-\alpha_3}\operatorname{adj}(A_2 - A_1) - \alpha_3(1-\alpha_3)\operatorname{adj}\Big[\Big(\frac{\alpha_1}{1-\alpha_3}A_1 + \frac{\alpha_2}{1-\alpha_3}A_2\Big) - A_3\Big], \tag{5.7}$$

which holds for any scalars α_i with $\alpha_1 + \alpha_2 + \alpha_3 = 1, \alpha_3 \neq 0,1$, and 3×3 matrices A_i. Applying this to $(5.6)_1$ and comparing it with $(5.6)_2$, using that $\operatorname{adj}(M_2 - M_1) = 0$, gives

$$\operatorname{adj}\Big[\frac{\tau_1}{1-\tau_3}M_1 + \frac{\tau_2}{1-\tau_3}M_2 - C\Big] = 0.$$

Hence,

$$C - \big[(1-\sigma)M_1 + \sigma M_2\big] = b \otimes m,$$

for some $b,m \in \mathbb{R}^3$ with $\sigma = \dfrac{\tau_2}{1-\tau_3} \in (0,1)$. Since $b \otimes m$ is not zero, by $(5.6)_{2,3}$, the result (5.3) follows.

In the rest of the discussion, we shall assume that $\lambda_3 = 0$ or 1. So we may write

$$\lambda_3 = \chi_{\Omega - E} \quad \text{and} \quad \lambda_1 + \lambda_2 = \chi_E \tag{5.8}$$

where E is the "martensite" set and $\Omega - E$ is the "austenite" set with respect to the reference configuration. We assume that E is not of full measure. The major issue here will be to show that (5.3) holds.

Step 2 The basic relation

From (5.4) and (5.8),

$$F(x) = \nabla y(x) = \chi_E(\lambda_1 M_1 + \lambda_2 M_2) + (1-\chi_E)C \quad \text{in } \Omega$$

so

$$F(x) - C = \chi_E(\lambda_1(M_1 - C) + \lambda_2(M_2 - C)) \quad \text{in } \Omega.$$

Introduce

$$z(x) = y(x) - Cx \quad \text{and} \quad N_i = M_i - C,\ i = 1,2. \tag{5.9}$$

With these notations,

$$\nabla z = \chi_E(\lambda_1 N_1 + \lambda_2 N_2) \quad \text{in } \Omega, \text{ where} \tag{5.10}$$

$$\lambda_1 + \lambda_2 = 1 \text{ in } E \qquad \text{and} \qquad N_2 = N_1 + a \otimes n . \tag{5.11}$$

By the hypothesis (5.2), rank $N_i \geq 2$.

Observe that since, by the equality of mixed second partials,

$$\int_\Omega \left(\frac{\partial z^i}{\partial x_\alpha} \frac{\partial \zeta}{\partial x_\beta} - \frac{\partial z^i}{\partial x_\beta} \frac{\partial \zeta}{\partial x_\alpha} \right) dx = 0 ,$$

we have that

$$N^i_{1\alpha} \int_\Omega \lambda_1 \frac{\partial \zeta}{\partial x_\beta} dx + N^i_{2\alpha} \int_\Omega \lambda_2 \frac{\partial \zeta}{\partial x_\beta} dx = N^i_{1\beta} \int_\Omega \lambda_1 \frac{\partial \zeta}{\partial x_\alpha} dx + N^i_{2\beta} \int_\Omega \lambda_2 \frac{\partial \zeta}{\partial x_\alpha} dx . \tag{5.12}$$

Let us set

$$\xi = \int_\Omega \lambda_1 \nabla \zeta \, dx \quad \text{and} \quad \eta = \int_\Omega \lambda_2 \nabla \zeta \, dx = \int_\Omega (\chi_E - \lambda_1) \nabla \zeta \, dx , \tag{5.13}$$

so (5.12) becomes our basic relation:

$$N^i_{1\alpha} \xi_\beta + N^i_{2\alpha} \eta_\beta = N^i_{1\beta} \xi_\alpha + N^i_{2\beta} \eta_\alpha , \tag{5.14}$$

or,

$$(c \cdot \xi) N_1 + (c \cdot \eta) N_2 = N_1 c \otimes \xi + N_2 c \otimes \eta , \quad \text{for any } c \in \mathbb{R}^3 . \tag{5.15}$$

Note that if $\xi = 0$, then

$$c \cdot \eta \, N_2 = N_2 c \otimes \eta ,$$

so N_2 is of rank one, unless $\eta = 0$. This is a contradiction. Similarly, $\eta = 0$ implies that $\xi = 0$. We conclude that

$$\xi = 0 \Leftrightarrow \eta = 0 \quad \text{and} \quad \xi + \eta \neq 0 \quad \text{for some } \zeta \in C^\infty_0(\Omega) . \tag{5.16}$$

The last assertion follows from the assumption that E is not of full measure.

Step 3 There are vectors $b, m \in \mathbb{R}^3$ with $b \otimes m$ unique and a unique $\sigma \in \mathbb{R}$ such that

$$N_1 = -\sigma a \otimes n + b \otimes m \quad \text{and} \qquad (5.17)$$
$$N_2 = (1-\sigma) a \otimes n + b \otimes m .$$

But we do not yet know in (5.17) that $0 < \sigma < 1$.

Case 1 Assume that $\xi \parallel \eta$ for some $\zeta \in C_0^\infty(\Omega)$. Write $\eta = \tau \xi$ for some real τ with $\tau \neq -1$ by (5.16). Setting this in (5.15) with c chosen such that $c \cdot \xi = 1$, we obtain

$$N_1 + \tau N_2 = (N_1 + \tau N_2) c \otimes \xi = b \otimes m \qquad (5.18)$$

is rank one. We may now solve (5.11) and (5.18) to deduce (5.17) with $\sigma = \dfrac{\tau}{1 + \tau}$ and

$$m = \int_\Omega \chi_E \nabla \zeta \, dx = \xi + \eta .$$

The choice of m is not unique; it may be any vector proportional to $\xi + \eta$.

Case 2 Suppose that ξ is not parallel to η. Choosing $c \in \mathbb{R}^3$ such that $c \cdot \xi = 1$ and $c \cdot \eta = 0$, then

$$N_1 = N_1 c \otimes \xi + N_2 c \otimes \eta = p \otimes \xi + q \otimes \eta . \qquad (5.19)$$

Similarly,

$$N_2 = N_1 c' \otimes \xi + N_2 c' \otimes \eta = q' \otimes \xi + r \otimes \eta . \qquad (5.20)$$

Substituting (5.19) and (5.20) into (5.15) yields that

$$(q - q') \otimes \eta \, (c \cdot \xi) = (q - q') \otimes \xi \, (c \cdot \eta) ,$$

whence $q = q'$. Since $N_2 - N_1 = a \otimes n$,

$$(q - p) \otimes \xi + (r - q) \otimes \eta = a \otimes n .$$

Since ξ is not parallel to η, it is necessarily the case that $q - p \parallel r - q \parallel a$. Let us write $q = p + ta$ and $r = q + sa = p + (s+t)a$ so

$$N_1 = p \otimes \xi + (p + ta) \otimes \eta \quad \text{and} \quad N_2 = (p + ta) \otimes \xi + (p + (s+t)a) \otimes \eta \quad (5.21)$$

Note that p is not parallel to a and $t \neq 0$ because rank $N_1 \geq 2$. Now

$$(1-\mu) N_1 + \mu N_2 = (p + \mu ta) \otimes \xi + (p + (\mu s + t)a) \otimes \eta$$

which is rank one if and only if $p + \mu ta = p + (\mu s + t)a$, whence

$$\mu = \frac{t}{t - s} \quad \text{and} \quad p + \mu ta = p + (\mu s + t)a = p + \frac{t^2}{t - s} a .$$

With this choice of μ,

$$(1-\mu) N_1 + \mu N_2 = \left(p + \frac{t^2}{t - s} a\right) \otimes (\xi + \eta) = b \otimes m . \quad (5.22)$$

From (5.11) and (5.22) we again obtain (5.17) with, for example,

$$m = \int_\Omega \chi_E \nabla\zeta \, dx = \xi + \eta .$$

To show uniqueness, suppose that for some σ, b, and m,

$$N_1 = -\sigma a \otimes n + b \otimes m .$$

Then

$$\operatorname{adj} (N_1 + \sigma a \otimes n) = 0 .$$

The left hand side of this equation is linear in σ with nonzero slope. Thus σ is uniquely determined and so is $b \otimes m$.

The implication of the uniqueness is that the mapping

$$\zeta \to \int_\Omega \chi_E \nabla\zeta \, dx = \xi + \eta \quad (5.23)$$

has a one dimensional range. Since the formula (5.23) provides us intuitively with the direction of the normal to E, we see that ∂E is in some sense planar.

Step 4 There are real valued Lipschitz functions of a single variable g and f such that

$$z(x) = g(x \cdot n)\, a + f(x \cdot m)\, b . \tag{5.24}$$

Substituting the reperesentation (5.17) into the formula (5.10), with $\lambda_2 = \lambda$ and $\lambda_1 = 1-\lambda$,

$$\nabla z = \chi_E (\lambda - \sigma)\, a \otimes n + \chi_E\, b \otimes m . \tag{5.25}$$

Choose a vector c with $c \cdot a = 0$ and $c \cdot b = 1$, possible since rank $N_1 > 1$. Then for $u(x) = c \cdot z(x)$,

$$\nabla u = \chi_E\, m ,$$

and hence, since Ω is convex, there is a Lipschitz function f(t) such that

$$u(x) = f(x \cdot m) .$$

In addition, we may write the set E as the cartesian product of two dimensional planes with an interval of the m - axis. More precisely, there is a (measurable) subset $I \subset \mathbb{R}$ such that

$$E = \{ x \in \Omega : x \cdot m = \alpha, \ \alpha \in I \} \tag{5.26}$$

Similarly, choosing $c' \in \mathbb{R}^3$, with $c' \cdot a = 1$ and $c' \cdot b = 0$, $w(x) = c' \cdot z(x)$ has the property

$$\nabla w = \chi_E (\lambda - \sigma)\, n$$

hence

$$w(x) = g(x \cdot n) .$$

Step 5 Equation (5.3) holds with $\sigma \in (0,1)$ and $\lambda(x) \equiv \sigma$ in a "propagation region."

Note that $\sigma \neq 0,1$ since N_i have rank two. The assumption that E is not of full measure means, by Fubini's Theorem for example, that there is a set

$$\Sigma = \{ x \in \Omega : x \cdot m = \alpha_o \} , \quad \text{for some } \alpha_o \notin I ,$$

such that almost every $x \in \Sigma$ is a Lebesgue point of ∇z and $\Omega - E$. Hence $\nabla z = 0$ in Σ by (5.25) and $z = \text{const.}$ in Σ. We may assume that $z = 0$ in Σ. Thus

$$0 = g(x \cdot n)\, a + f(\alpha_0)\, b \quad \text{in } \Sigma .$$

Since a and b are independent,

$$g(x{\cdot}n) = f(\alpha_0) = 0 \qquad \text{in } \Sigma . \tag{5.27}$$

Let us set

$$D = \{ x \in \Omega\text{: } x \text{ lies on a plane } x \cdot n = \beta \text{ which intersects the plane } \Sigma \}, \tag{5.28}$$

which we call a propagation region for Σ. From (5.27) we have that $g(x{\cdot}n) = 0$ in D, hence

$$\nabla g = (\lambda - \sigma)\,\chi_E = 0 \qquad \text{in } D .$$

Thus $\lambda = \sigma$ in $E \cap D$ a.e., so (i) $\sigma \in (0,1)$ and (ii) $\lambda = \sigma = \text{const.}$ in $E \cap D$ a.e.

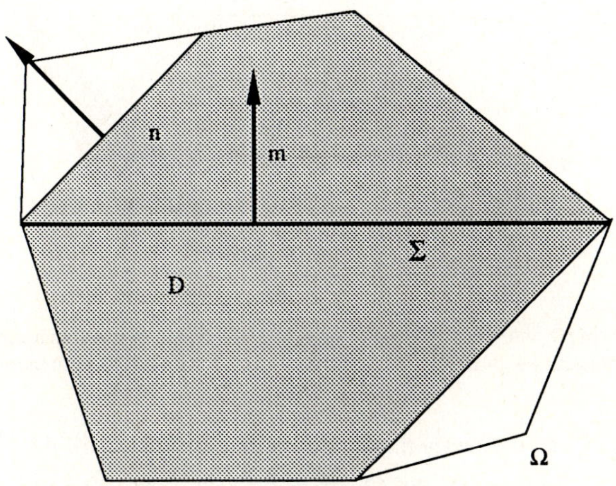

Fig. 10 Sketch of the propagation region D of Σ.

Inside the propagation region of Σ, there may be a mix of martensite and austenite but the martensite occurs with fixed variant proportion σ, cf. Fig. 10. Outside the propagation region of

Σ, there may be pure martensite, if such a domain does not intersect with any other propagation region, possibly of varying proportions, but usually a single variant, cf. Fig. 11. So, for example, under the assumption (5.8), we may write that the Young measure (5.4)

$$\nu_x = \chi_E((1-\sigma)\,\delta_{M_1} + \sigma\,\delta_{M_2}) + (1-\chi_E)\,\delta_C, \quad x \in D. \tag{5.29}$$

An analogous representation of the Young measure may be given under the assumptions of Step 1. In this case,

$$\lambda \equiv \lambda_3 = \lambda_3(x \cdot m) \qquad \text{and}$$

$$\nu_x = (1-\lambda)\,((1-\sigma)\,\delta_{M_1} + \sigma\,\delta_{M_2}) + \lambda\,\delta_C, \quad x \in D,$$

where the propagation region is determined in a fashion analogous to (5.28). We have not observed this configuration in the laboratory, but it remains a mathematical possibility.

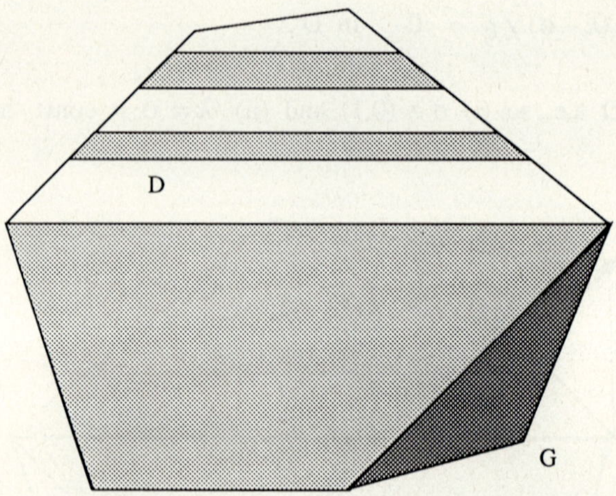

Fig. 11 In D there is a mixture of austenite and finely twinned martensite and the latter occurs with fixed variant proportions. In G there is pure martensite.

The condition (5.3) is exactly the condition that is the basis of the crystallographic theory of martensite presented, for example, by Christian [16]. For the origins of the theory, see Wechsler, Lieberman, and Read [73]. When the matrices M_i and C are only restricted to lie in potential wells, (5.3) gives rise to an algebraic problem: Given $R \in SO(3)$ with $RU_2 - U_1 = a \otimes n$, find $Q \in SO(3)$, $b, m \in \mathbb{R}^3$, and $\sigma \in (0,1)$ such that

$$Q - [\,\sigma R U_2 + (1-\sigma) U_1\,] = b \otimes m. \tag{5.30}$$

An existence theorem for (5.30) is given in [6], not restricted to the cubic/tetragonal transformation. To summarize briefly what is known in the cubic/tetragonal case, m, the reference normal to the austenite/martensite interface, turns out irrational but near a member of the family of {110} planes, if U_1 is near $\mathbb{1}$, and the angle between m and n is nearly 60°, both of which are observed experimentally. Although for one kind of microstructure, our approach and the Crystallographic Theory of Martensite are in complete agreement, the crystallographic theory is not based on energy considerations but on a kinematic condition known as the *double shear mechanism*. We think that a distinct advantage of our approach is that we are able to discuss general microstructures. To explore its general usefulness, additional predictions of the theory which may be made subject to experimental test are described in [7].

The examples of the preceding two sections illustrate how the behavior of minimizers of a variational principle may be reduced to questions of an algebraic character by the introduction of an appropriate parametrized measure. In addition to the papers already cited, ideas used in such analyses may be found in Matos [50] and Pedregal [56].

Acknowledgements *We wish to thank our collaborators John Ball and Michel Chipot. We also wish to acknowledge many conversations with Jerry Ericksen and Mitchell Luskin.*

References

1. Ball, J. M. 1976 Constitutive equations and existence theorems in nonlinear elastostatics, *Heriot-Watt Symp., I*, (R. Knops,ed.) Pitman, 187-241

2. Ball, J. M. 1980 Strict convexity, strong ellipticity, and regularity in the calculus of variations, Math Proc Camb Phil Soc, 87, 501-513

3. Ball, J. M. 1984 Singular minimizers and their significance in elasticity, *Phase Transformations and Material Instabilities in Solids,* (Gurtin, M., ed) Academic Press, 1-20

4. Ball, J. M. 1989 Sets of gradients with no rank-one connections (to appear)

5. Ball, J. M. 1989 A version of the fundamental theorem for Young measures, these proceedings

6. Ball, J. M. and James, R. 1987 Fine phase mixtures as minimizers of energy, Arch. Rat. Mech. Anal.,100, 15-52

7. Ball, J. M. and James, R. 1989 Proposed experimental tests of a theory of fine microstructure and the two well problem (to appear)

8. Ball, J. M. and Murat, F. 1984 $W^{1,p}$ - quasiconvexity and variational problems for multiple integrals, J. Fnal Anal, 58,225-253

9. Ball, J. M., and Knowles, G. 1986 Liapunov functions for thermomechanics with spatially varying boundary temperatures, Arch. Rat. mech. Anal., 92, 193-204

10. Basinski, Z. S. and Christian, J. W. 1954 Experiments on the martensitic transformations in single crystals of indium-thallium alloys, Acta Met. 2, 149-166

11. Battacharya, K 1989 Thesis, University of Minnesota

12. Burkart, M. W. and Read, T. A. 1953 Diffusionless phase change in the indium-thallium system, J. of Metals, 5, 1516-1524

13. Chipot, M. and Evans, L. C. 1986 Linearisation at infinity and Lipschitz estimates for certain problems in the calculus of variations, Proc. R. Soc. Edin., 102A, 291-303

14. Chipot, M. and Kinderlehrer, D. 1988 Equilibrium configurations of crystals, Arch. Rat. Mech. Anal. 103, 237-277

15. Chipot, M., Kinderlehrer, D., and Vergara-Caffarelli, G. 1986 Smoothness of linear laminates, Arch. Rat. Mech. and Anal. 96, 81-96

16. Christian, J. W. 1975 *The theory of transformations in metals and alloys*, Pergamon

17. Collins, C. and Luskin, M. 1989 The computation of the austenitic-martensitic phase transition (these proceedings)

18. Dacorogna, B. 1982 *Weak continuity and weak lower semicontinuity of nonlinear functionals*, Springer Lecture Notes 922 (1982)

19. Dacorogna, B. and Fusco, N. 1985 Semi-continuité des fonctionnelles avec contraintes du type "det grad u > 0", Boll. U M I, 6, 179-189

20. Eftis, J., MacDonald, D. E., and Arkilic, G. M. 1971 Theoretical calculation of the pressure variation of second-order elastic coefficients for alkali metals, Mater. Sci. Eng. 7, 141-150

21. Ericksen, J. L. 1973 Loading devices and stability of equilibrium, in Nonlinear Elasticity, Academic Press, 161-173

22. Ericksen, J. L. 1977 Special topics in elastostatics, Adv. in appl. mechanics, (C.-S. Yih, ed.) Academic Press 7, 189-243

23. Ericksen, J. L. 1979 On the symmetry of deformable crystals, Arch. Rat. Mech. Anal. 72, 1-13

24. Ericksen, J. L. 1980 Some phase transitions in crystals, Arch. Rat. Mech. Anal. 73, 99-124

25. Ericksen, J. L. 1981 Some simpler cases of the Gibbs phenomenon for thermoelastic solids, J.of thermal stresses, 4 , 13-30

26. Ericksen, J. L. 1983 Ill posed problems in thermoelasticity theory, *Systems of Nonlinear Partial Differential Equations*, (Ball, J., ed) D. Reidel, 71-95

27. Ericksen, J. L. 1984 The Cauchy and Born hypotheses for crystals, *Phase Transformations and Material Instabilities in Solids*, (Gurtin, M., ed) Academic Press, 61-78

28. Ericksen, J. L. 1986 Stable equilibrium configurations of elastic crystals, Arch. Rat. Mech. Anal. 94, 1-14

29. Ericksen, J. L. 1987 Twinning of crystals I, *Metastability and Incompletely Posed Problems*, IMA Vol. Math. Appl. 3,(Antman, S., Ericksen, J.L., Kinderlehrer, D., Müller, I.,eds) Springer, 77-96

30. Ericksen, J. L. 1988 Some constrained elastic crystals, *Proc. Symp. Material instabilities in continuum mechanics, Heriot-Watt* (Ball, J. M. ed.) Oxford, 119-136

31. Fonseca, I. 1985 Variational methods for elastic crystals, Arch. Rat. Mech. Anal., 97, 189-220

32. Fonseca, I. 1988 The lower quasiconvex envelope of the stored energy function for an elastic crystal, J. Math. pures et appl, 67, 175-195

33. Fonseca, I. 1989 Interfacial energy and the Maxwell rule, Arch. Rat. Mech. Anal.

34. Gibbs, J. W. 1961 On the equilibrium of heterogeneous substances, *The Scientific Papers of J. Willard Gibbs,* Vol 1., Dover

35. Gurtin, M. E. 1983 Two phase deformations of elastic solids, Arch. Rat. Mech. Anal., 84, 1-29

36. Gurtin, M. E. 1986 On phase transitions with bulk, interfacial, and boundary energy, Arch. Rat. Mech. Anal., 96, 243-264

37. Herring, C. 1951 Some theorems on the free energy of crystal surfaces, Phys. Rev. 82, 87-93

38. James, R. D. 1986 Displacive phase transformations in solids, J. Mech. Phys. Solids, 34, 359-394

39. James, R. D. 1986 Phase transformations and non-elliptic free energy, *New Perspectives in Thermodynamics* (Serrin, J., ed) Springer, 223-239

40. James, R. D. 1987 The stability and metastability of quartz, *Metastability and Incompletely Posed Problems,* IMA Vol. Math. Appl. 3,(Antman, S., Ericksen, J.L., Kinderlehrer, D., Müller, I.,eds) Springer, 147-176

41. James, R. D. 1988 Microstructure and weak convergence, *Proc. Symp. Material instabilities in continuum mechanics,* Heriot-Watt (Ball, J. M. ed.) Oxford, 175-196

42. Kinderlehrer, D. 1987 Twinning in crystals II, *Metastability and Incompletely Posed Problems,* IMA Vol. Math. Appl. 3,(Antman, S., Ericksen, J.L., Kinderlehrer, D., Müller, I.,eds) Springer, 185-211

43. Kinderlehrer, D. 1988 Remarks about the equilibrium configurations of crystals, *Proc. Symp. Material instabilities in continuum mechanics,* Heriot-Watt (Ball, J. M. ed.) Oxford, 217-242

44. Kinderlehrer, D. and Vergara-Caffarelli, G. The relaxation of functionals with surface energies, to appear

45. Kohn, R. V. and Strang, G. 1987 Optimal design and relaxation of variational problems, I,CPAM 34, 113-137

46. Kohn, R. V. and Strang, G. 1987 Optimal design and relaxation of variational problems, II,CPAM 34, 139-182

47. Kohn, R. V. and Strang, G. 1987 Optimal design and relaxation of variational problems, III,CPAM 34, 353-377

48. Marcellini, P. 1986 On the definition and the lower semicontinuity of certain quasiconvex integrals, Ann. Inst. H. Poincaré; Analyse non linéaire, 3, 391-409

49. Mascolo, E. and Schianchi, R. 1983 Existence theorems for non convex problems, J. Math pures et appl, 62,(1983), 349-359

50. Matos, J. 1989 Thesis, University of Minnesota

51. Morrey, C. B., Jr. 1966 *Multiple Integrals in the Calculus of Variations,* Springer

52. Müller, I. 1979 A model for a body with shape memory, Arch. Rat. Mech. Anal., 70, 61-77

53. Müller, I. and Wilmanski, K. 1980 A model for phase transition in pseudoelastic bodies, Il Nuovo Cim., 57B, 283-318

54. Parry, G. 1981 On phase transitions involving internal strain, Int. J. Solids Structures, 17, 361-378

55. Parry, G. On shear bands in unloaded crystals, to appear

56. Pedregal, P. Thesis, University of Minnesota

57. Pego, R. 1985 Phase transitions in one dimensional nonlinear viscoelasticity: admissibility and stability, Arch. Rat. Mech. Anal. 97, 353-394

58. Pego, R. 1987 Phase transitions in one dimensional nonlinear viscoelasticity: admissibility and stability, *Dynamical Problems in Continuum Physics,* I M A Vol. Math. Appl. 4 (Bona, J., Dafermos, C., Ericksen, J. L., and Kinderlehrer, ., eds)277-288

59. Pipkin, A. C. 1986 The relaxed energy density for isotropic elastic membranes, IMA J Appl Math, 36, 85-99

60. Pippard, A. B. 1957 The elements of classical thermodynamics, Cambridge

61. Pitteri, M. 1985 On v + 1 lattices, J. Elasticity, 15, 3-25

62. Pitteri, M. 1986 On type-2 twins in crystals, Int. J. Plasticity, 2, 99-106

63. Pitteri, M. 1987 A contribution to the description of natural states for elastic crystalline solids, *Metastability and Incompletely Posed Problems,* IMA Vol. Math. Appl. 3,(Antman, S., Ericksen, J.L., Kinderlehrer, D., Müller, I.,eds) Springer, 295-310

64. Reshetnyak, Yu. G. 1967 Liouville's theorem on conformal mappings under minimal regularity assumptions, Siberian Math J., 8, 631-653

65. Reshetnyak, Yu. G. 1967 On the stability of conformal mappings in multidimensional spaces, Siberian Math J., 8, 69-85

66. Simpson, H. C. and Spector, S. J. 1987 On the sign of the second variation in finite elasticity, Arch. Rat. Mech. Anal. 98, 1-30

67. Slemrod, M. 1986 Interrelationships among mechanics, numerical analysis, compensated compactness, and oscillation theory, *Oscillation Theory, Computation, and Methods of Compensated Compactness,* I M A Vol. Math. Appl. 2, (Dafermos, C., Ericksen, J. L, 309-336. , Kinderlehrer, D., and Slemrod, M. eds)

68. Tartar, L. 1979 Compensated compactness and applications to partial differential equations, *Nonlinear analysis and mechanics: Heriot-Watt symp. IV,* (Knops, R. J. ed) Pitman, (1979), 136-212

69. Tartar, L. 1983 The compensated compactness method applied to systems of conservation laws, *Systems of nonlinear partial differential equations* (Ball, J. M., ed) Riedel, 263-285

70. Tartar, L. 1984 Étude des oscillations dans les équations aux dérivées partielles nonlinéaires, Springer Lect. Notes Physics, 195, 384-412

71. Tisza, L. 1951 On the general theory of phase transformations in solids, (Smoluchowski, R., Mayer, J. E., and Weyl, W. A., eds) Wiley, 1-35

72. van Tendeloo, G., van Landuyt, J., and Amelinckx, S. 1976 The $\alpha-\beta$ phase transition in quartz and AlPo4 as studied by electron microscopy and diffraction, Phys. Stat. Sol., a33, 723-735

73. Wechsler, M. S., Lieberman, D. S., and Read, T. A. 1953 On the theory of the formation of martensite, Trans. AIME J. Metals, 197, 1503-1515

74. Young, L. C. 1969 *Lectures on calculus of variations and optimal control theory,* W.B. Saunders

INVARIANCE PROPERTIES OF INVISCID FLUIDS OF GRADE N

Pierre CASAL and **Henri GOUIN**

P.C. : Université de Provence, place Victor-Hugo, 13331 Marseille Cedex 3, France.

H.G. : Faculté des Sciences et Techniques, rue Henri Poincaré, 13397 Marseille Cedex 13, France.

Fluids of grade n are continuous models for dynamic changes of phase which avoid the surfaces of discontinuity representing the capillary layers in liquid-vapour interfaces. We recall the thermodynamic form of the equation of motion for inviscid fluids of grade n [1]. First integrals and circulation theorems are deduced and a general classification of flows is given.

1. Introduction

In continuum mechanics, first gradient media cannot represent fluids with strong density variations. Material surfaces need their own characteristic behaviour and energy properties [2].

Fluids of grade n (n is any integer) are continuous media with an internal energy per unit mass which is a function of entropy s, density ρ and spatial gradients up to n-1 order:

$$\varepsilon = \varepsilon \left(s, \text{grad } s, \ldots, (\text{grad})^{(n-1)} s, \rho, \text{grad } \rho, \ldots, (\text{grad})^{(n-1)} \rho \right)$$

D.J. Korteweg pointed out the advantage of fluids of grade n with respect to the density and removed the discontinuity surfaces representing the capillary layers in liquid-vapour interfaces [3].

We improve the model accuracy by taking into account the successive gradients of density. Recently, this approach was used in the case of dynamic changes of phase [4]. Till now, the thermodynamic part of the fluid was neglected. In fact, it is difficult to take into account the gradients of temperature (it is not possible to consider a virtual displacement of temperature), and it is necessary to use the entropy by way of the internal energy.

The equation of motion with strong gradients of density and entropy is obtained. One can accord importance to the fact that the thermodynamic form of the equation of motion is independent of the order n of the fluid. Flows with strong variations of temperature such as those associated with combustion phenomena and non-isothermal interfaces may be considered.

The conservative flows of perfect fluids are only a mathematical model. So it is with fluids of grade n. However, the study of their structure is fundamentally necessary. The conclusions for fluids of grade n are of the same kind as for compressible fluid [5]: it is possible to obtain a Clebsh transformation of the equation of motion (*potential equations*) and many invariance properties and first integrals. The potential equations may be used to classify motions in the same manner as in the case of perfect fluids [6,7].

Recall that the simplest case is thermocapillary fluids: the internal energy is a function only of density, entropy and their spatial derivatives ($\varepsilon = \varepsilon$ (s, grad s, ρ, grad ρ)) and $n > 2$ adds little except for complications in the momentum conservation equation. In considering flows through liquid-vapour interfaces and using convenient physical units, equations of thermocapillary fluids may study non-isothermal motions and give an interpretation of film boiling phenomena [8,9].

2. Thermodynamic form of the equation of motion for inviscid fluids of grade n [1]

The principle of virtual works allows to research the equation of motion. In the case of a perfect fluid motion, it is written with Hamilton's principle [5,6,10].

Variations of motion of particles are deduced from

$$X = \Psi(x, t, \alpha)$$

where α is a parameter defined in a neighbourhood of zero; it is associated with a family of virtual motions of the fluid [5]. The real motion corresponds to $\alpha = 0$.

A particle is represented, in Lagrangian coordinates, by the position X (X^1, X^2, X^3) it occupies in a reference space D_o. At time t, its position is given, in D_t, by the eulerian representation x (x^1, x^2, x^3) (see appendix 1).

Virtual displacements associated with any variation of real motion can be written:

$$\delta X = (\partial\Psi/\partial\alpha)\,|_{\alpha=0}$$

This variation is dual with Serrin's [5]: they are mathematically equivalent [11]. The Lagrangian of a fluid of grad n is:

$$L = \rho\,(1/2\, V^*V - \varepsilon - \Omega)$$

V denotes the velocity vector, Ω the extraneous force potential defined in D_t and * the transposition in D_t.

Between times t_1 and time t_2, hamiltonian action is written:

$$a = \int_{t_1}^{t_2} \int_{D_t} L \, dv \, dt$$

The density verifies the balance equation

$$\rho \det F = \rho_o(X)$$

where ρ_o is defined in D_o and F is the deformation gradient.

We deduce: $$\frac{d\rho}{dt} = - \rho \operatorname{div} V \qquad (1)$$

The variation of entropy is the sum of a variation associated with the virtual motion and another one, $\delta_1 s$, related to the particle [9]:

$$\delta s = (\partial s/\partial X)\,.\ \delta X + \delta_1 s\,,$$

where $\delta_1 s$ and δx are independent. To obtain the equation of motion, we consider the case $\delta_1 s = 0$. Classical methods of variational calculus provide the variation of hamiltonian action:

$$\delta a = 0$$

A calculus like [1] yields:

$$\delta a = \int_{t_1}^{t_2} \int_{D_t} \rho_o \, [(d/dt(V^* F) - \theta \, grad_o^* s - grad_o^* m \,] \, \delta X \; dv_o \, dt$$

where
$$m = 1/2 \, V^* V - h - \Omega$$

$$h = \varepsilon + p/\rho$$

(F)

$$p = \rho^2 \varepsilon'_\rho + \rho \sum_{k=1}^{n-1} (-1)^k \, div^{(k)}(\rho \, \partial\varepsilon/\partial grad^{(k)}\rho)$$

$$\theta = \varepsilon'_s + (1/\rho) \sum_{k=1}^{n-1} (-1)^k \, div^{(k)} (\rho \, \partial\varepsilon/\partial grad^{(k)} s)$$

In this calculus, we have taken into account that $\delta\rho = \rho \, div_o (\delta X) + (1/\det F)(\partial\rho_o/\partial X) \, \delta X$ and $\delta V = - F \, d/dt(\delta X)$.
$div^{(k)}$ and $grad^{(k)}$ are divergence and gradient operators in D_t reiterated k-times. div_o and $grad_o$ are divergence and gradient operators in D_o.

Obviously, p, θ, h have the same dimensions as a pressure, a temperature and a specific enthalpy do. We name them pressure, temperature and enthalpy of fluid of grade n.

Writing that for any displacement δX zero on the edge of D_o, $\delta a = 0$, we get the equation:

$$d/dt \, (V^* F) - \theta \, grad_o^* s - grad_o^* m = 0.$$

Noting that $(\Gamma^* + V^* \partial V/\partial x)F = d/dt(V^* F)$, we obtain:

$$\Gamma = \theta \, grad \, s - grad(\, h + \Omega \,) \tag{2}$$

In case of inviscid thermocapillary fluids, we prove that equations of motion and energy imply that [12]:

$$\rho \, \theta \frac{ds}{dt} + div \, q - r = 0 \tag{3}$$

where q is the heat flux vector and r the heat supply (the dissipation function is zero). This result is extended to fluids of grade n [13].

A thermodynamic behaviour must be added to these equations. It gives us the heat flux vector q: for example, the Fourier law is :

$$q = -k \, grad \, \theta \quad et \quad k \geq 0 \tag{4}$$

If the coefficient of conductivity is zero, the fluid is perfect. The heat supply is assumed to be given in similar fashion to the extraneous force potential.

With the heat conduction inequality, we obtain the Clausius-Duhem inequality and fluids of grade n are compatible with the second law of thermodynamics [14,15]:

$$\rho \frac{ds}{dt} + div \frac{q}{\theta} - \frac{r}{\theta} \geq 0$$

So, motions of inviscid fluids of grade n assume the system:

$$\begin{gathered}\Gamma = \theta \text{ grad } s - \text{grad}(h + \Omega) \\ \frac{\partial \rho}{\partial t} + \text{div}\, \rho V = 0 \\ \rho\, \theta \frac{ds}{dt} + \text{div}\, q - r = 0 \\ q = -k \text{ grad } \theta \quad \text{et} \quad k \geq 0\end{gathered} \tag{A}$$

The thermodynamic form of equations for inviscid fluid are equivalent to the classic balance equations. But, we must notice that these equations are of the same form for all fluids (independently of the order n). Nevertheless, the balance equations become considerably complicated when n increases. In our way, only expressions for temperature and enthalpy become more complicated (and we are obviously led to rectify the definitions[1] of these two terms).

3. Thermodynamic hypothesis on the motion

If we study the possible invariance properties of the motion, we must first assume that the heat supply allows a thermodynamic behaviour just expressing that a pure thermodynamic quantity T (function only of temperature and entropy) has a zero material derivative or is constant everywhere in the flow. We assume that T remains constant on every trajectory (so it has a zero material derivative) or - a stronger assumption - that T is constant everywhere in the flow. Then, equations (3) and (4) are replaced by:

$$\frac{dT}{dt} = 0 \qquad (5^1) \qquad (5)$$

or

$$T = \text{const.} \qquad (5^2)$$

For example, we will be considering the case of an isentropic motion:

$$T = s \quad \text{and} \quad \frac{ds}{dt} = 0,$$

and the case of an isothermal motion:

$$T = \theta \quad \text{and} \quad \theta = \text{const.}$$

They are two limit cases: the first one is associated with fast motions, the second one with slow motions.[2] Then, the fluid motions satisfy the system consisting of equations (1), (2) and (5).

Hypothesis: We assume that $T = f(\theta, s)$ is differentiable and $\partial T/\partial\theta$ is not zero. Let $H(T,s)$ be a function twice differentiable of the variables T and s such that $\partial H/\partial s = \theta(T,s)$. H is defined by its differential with an unknown function of T:

---

[1] These definitions are given only by relation (F).

[2] The Euler equation represents the equation of motion for classical inviscid fluids. Such motions may be obtained with non-negligible heat flux and heat supply. They correspond to physical situations. As a matter of fact, in the non-dimensional equation of motion for viscous fluids [16,17], it is often justified to neglect terms due to viscosity, when, with respect to other physical quantities, the Reynolds number is large. Nevertheless, for large temperature gradients and with a Prandlt number of order one, it is not possible to remove the terms of thermal conductivity. Such a case arises when one studies the "thermal" boundary layer [18,19] and for some slow motions of natural convection in fluid with gravity and heat sources associated with large differences of temperature [20].

$$dH = \theta\, ds - u\, dT$$

The two variables u and T will be used instead of θ and s for studying the general case by transforming equation (2) and using the method of appendix 2. Let us note $g = h - H$, equation (2) is written:

$$\Gamma = u \text{ grad } T - \text{grad}(g + \Omega) \tag{6}$$

Notice that: if $T = \theta$, g is the free enthalpy $h - \theta s$ and $u = -s$

Motions satisfy system (B) formed with equations (1), (5) and (6).

By means of first integrals (research of scalars, functions of time and state of the fluid (density, entropy, velocity with a zero material derivative)), this common expression allows us to make a single classification for all fluids.

The momentum conservation equation expresses the balance of forces operating on every part of the fluid (inertial forces, body forces, stresses).

System (B) is not in balance, but expresses the evolution of the fluid: The equation $\frac{dT}{dt} = 0$ indicates that T is constant along a trajectory and the continuity equation $\frac{d\rho}{dt} = - \rho \text{ div } V$ yields the density variation. It will be shown that equations (2) or (6) describe the evolution of the velocity field.

First, the notion of convective derivative is recalled.

4. Material derivative and convective derivative (appendix 1)

*The **material derivative** of any tensor of the physical space is its derivative with respect to time, when one follows the particle along its motion. This tensor has an image in the reference space. The derivative with respect to time of this image has for reciprocal image the **convective derivative of this tensor which we denote by** d_c.*

These two derivatives are the same for a scalar tensor and a tensor is convected by the stream if its convective derivative is zero.

Transposing equation (6), we obtain the convective derivative of the velocity covector:

$$d_c V^* = \text{grad}^*(\tfrac{1}{2} V^2 - g - \Omega) - u \text{ grad}^* T \tag{7}$$

This equation, equivalent to equation (6) directly yields the Helmholtz-Kelvin theorems in their most general form.

Let us consider an arbitrary fluid curve C and $I(t) = \int_C V^* dx$ the circulation of the convective derivative, consequently,

$$\frac{dI}{dt} = \int_C d_c V^* dx = \int_C d(\tfrac{1}{2} V^2 - g - \Omega) + \int_C u\, dT$$

If C is a closed curve, the first integral is zero.

If, moreover, C is located on a surface where T = const. , which is not conflicting since those surfaces are themselves fluid, the second integral is zero also and I is constant.

The proper Kelvin theorem corresponds to the particular case of *homentropic* motions for which entropy is constant everywhere in the fluid (T = s = const.). The circulation of the velocity field on any closed curve is constant.

Due to Noether's theorem, it is known that any conservation law can be represented by an invariant group. It has been shown that the conservation law expressed by the Kelvin theorems associated with isentropic fluid curves corresponds to the group of permutations consisting of particles of equal entropy [21,22]. It is clear that this group keeps the equations of motion invariant for both a classical perfect fluid and a fluid of grade n. It is tempting to define a general perfect fluid by identifying it with this invariant group or consequently, with a continuous medium whose motions verify the Kelvin theorems.

5. Potential equations

Let φ and ψ be two scalar fields with material derivatives such that:

$$\frac{d\varphi}{dt} = \frac{1}{2}V^2 - g - \Omega \quad \text{et} \quad \frac{d\psi}{dt} = u$$

Equation (7) is equivalent to:

$$d_c V^* = d_c(\text{grad}^*\,\varphi + \psi\,\text{grad}^* T)$$

The covector V^* and $W^* = \text{grad}^*\,\varphi + \psi\,\text{grad}^*\,T$ have the same convective derivative. So, their difference is a covector convected by the stream. This leads to the system:

$$V = \text{grad}\,\varphi + \psi\,\text{grad}\,T + C \tag{C}$$

$$\frac{dT}{dt} = 0\,, \quad \frac{d\varphi}{dt} = \frac{1}{2}V^2 - g - \Omega\,, \quad \frac{d\psi}{dt} = u\,, \quad d_c\,C^* = 0$$

System (B) is equivalent to system {(C) + (1)}.

The Helmholtz theorem generalized above is reduced to: the circulation of C along any fluid curve is constant.

Transforming the integral with the Stokes formulae, we deduce that vector $U = 1/\rho\ \text{rot}\,C$ is convected by the stream ($d_c\,U = 0$) i.e. :

$$\frac{dU}{dt} - \frac{\partial V}{\partial x}\,U = 0 \tag{8}$$

It is possible (see appendix 3), to replace vector C by two scalars γ and ν *convected by the stream* (gradφ, grad ψ and grad ν independent).

System (C) transforms into system (D) ***of potential equations :***

$$V = \text{grad}\,\varphi + \psi\,\text{grad}\,T + \gamma\,\text{grad}\,\nu \tag{D}$$

$$\frac{dT}{dt} = 0, \quad \frac{d\varphi}{dt} = \frac{1}{2}V^2 - g - \Omega, \quad \frac{d\psi}{dt} = u\,, \quad \frac{d\gamma}{dt} = 0, \quad \frac{d\nu}{dt} = 0\,.$$

In the general case, the velocity field is expressed with five potentials and T is one of them[1].

6. Classification of motions

As in the case of T, potentials γ and ν have a zero material derivative. They are first integrals of the motion [23].

If these special potentials - which have constant values on every trajectory - have a constant value everywhere in a region of the fluid, they keep this constant value at later time.

In this region, which is convected by the flow, the fluid motion will be described such that the velocity will be expressed by a least number of potentials: that leads to characteristic kinematic properties.

In an adjoining region, the motion may be different. Then the flow is divided into different domains which move, remain adjacent and keep their own characteristics.

In hydrodynamics (incompressible perfect fluids), it is well known that the flow is divided into vortical and irrotational areas which do not mix because of the absence of viscosity.

The following study shows that, for all inviscid fluids (compressible or not) and even for fluids of any grade, the vortical area may be divided into parts of particular kinds[2].

In fact, we have a *scalar first integral* T and a *first integral with a covector nature*. The classical notion of first integral with a constant value must be replaced for a non-scalar tensor, by the notion of a value *whose image is constant in reference space,* i.e. convected by the stream or with a zero convective derivative.

We obtain different types of flow when C is zero or T is constant.

OLIGOTROPIC MOTIONS

If the covector C* is zero (either $\gamma = 0$, or $\nu = \text{const.}$), the velocity field has a particular kinematic type:

a) It can be expressed with the use of three potentials instead of five and the equations of motion are:

$$V = \text{grad}\ \varphi + \psi\ \text{grad}\ T \tag{9}$$

$$\frac{dT}{dt} = 0, \quad \frac{d\varphi}{dt} = \frac{1}{2}V^2 - g - \Omega \quad , \quad \frac{d\psi}{dt} = u\,,$$

b) Applying the curl to the two sides of equation (9),

$$\text{rot}\ V = \text{grad}\ \psi \wedge \text{grad}\ T$$

then,

$$\text{grad}^*T\ .\ \text{rot}\ V = 0 \tag{10}$$

Vortex lines are located on the surfaces $T = \text{const.}$

c) This property is characteristic. Equation (10) is written without utilizing potentials φ et ψ.

---

[1]The velocity field also depends on the density which is not considered as a potential.

[2]Let us note that systems (A) and (B) which are equivalent, assume that the flow is not crossed by shock waves. A shock wave generally transforms the nature of flow. This study has been done by Gouin [11,24]. This is not the subject of this paper but we need to mention it.

d) These motions are intermediate between irrotational motions and general vortical ones: this is one reason for calling them *oligotropic* .

e) If at an instant, the velocity field verifies relation (10), it will be true at any moment and the motion is oligotropic.

MOTIONS WHEN T IS CONSTANT (T = s : HOMENTROPIC MOTIONS)

If the thermodynamic relation is T = const. (which is stronger than $\frac{dT}{dt} = 0$), the velocity field has an another particular type. (For example, when T=s , the motion is not only isentropic but also homentropic).

a) It can also be expressed with three scalar potentials

$$V = \text{grad}\ \varphi + \gamma\ \text{grad}\ \nu$$

$$\frac{dT}{dt} = 0, \quad \frac{d\varphi}{dt} = \frac{1}{2}V^2 - g - \Omega\ , \quad \frac{d\gamma}{dt} = 0\ , \quad \frac{d\nu}{dt} = 0$$

but Clebsch's representation [16] is practically worthless because curves γ = const. and ν = const. are precisely the vortical lines whose complexity is well known [25].

It is easiest to write these equations in an equivalent form:

$$V = \text{grad}\ \varphi + C$$

$$\frac{dT}{dt} = 0\ , \quad \frac{d\varphi}{dt} = \frac{1}{2}V^2 - g - \Omega\ , \quad d_C\, C = 0$$

b) From (10) we deduce that vector $\frac{1}{\rho}$ rot V is a convected vector :

$$d_C\ \left(\frac{1}{\rho}\ \text{rot}\ V\right) = 0 \qquad (11)$$

This is Cauchy's theorem for homentropic motions, which is generalized at any moment for which a thermodynamic potential T is constant in the flow.

c) Equation (11) is characteristic of these motions.

IRROTATIONAL MOTIONS

If an oligotropic motion is also a motion with T constant, the velocity field is derived from only one potential :

$$V = \text{grad}\ \varphi\ , \quad \frac{d\varphi}{dt} = \frac{1}{2}V^2 - g - \Omega\ .$$

Of course, this leads to rot V = 0 at any instant[1].

This is the well known case of irrotational motions: the velocity field is usually the gradient of a single potential whose evolution is given by the Bernoulli equation.

---

[1] It is to be noticed that if, at an instant rot V = 0 everywhere in the fluid, it is not necessarily true at later times, as it is the case in the hydrodynamics of compressible fluids.

Permanent motions

For a permanent motion, Equation (6) may be written on a generalized Crocco-Vaszonyi form:

$$\text{rot}\, V \wedge V = u \,\text{grad}\, T - \text{grad}\, G \qquad (12)$$

where $G = \frac{1}{2}V^2 + g + \Omega$.

If T = s, g is the enthalpy and G the Lamb function [16]. (surfaces where G is constant are Lamb surfaces).

It follows: $\frac{dT}{dt} = 0 \Leftrightarrow \frac{dG}{dt} = 0$

G is a scalar first integral of the motion and we take it instead of the function ν which could be an arbitrary choice.

Then, potential equations for general permanent flows are:

$$V = \text{grad}\, \varphi_1 + \psi_1 \,\text{grad}\, T + \gamma_1 \,\text{grad}\, G$$

$$\frac{d\varphi_1}{dt} = \frac{1}{2}V^2 - g - \Omega\,, \quad \frac{d\psi_1}{dt} = u\,, \quad \frac{dT}{dt} = 0\,, \quad \frac{d\gamma_1}{dt} = 0, \quad \frac{dG}{dt} = 0$$

But $\frac{dT}{dt} = 0$ and $\frac{dG}{dt} = 0 \Rightarrow \frac{\partial \varphi_1}{\partial t} = -G$. If $\varphi = \varphi_1 + tG$, φ is independent of time.

Let us denote $\gamma_1 = \gamma + t$ and $\psi_1 = \psi$, then, the equations of motion may be written with potentials which all are independent of time:

$$V = \text{grad}\, \varphi + \psi \,\text{grad}\, T + \gamma \,\text{grad}\, G \qquad (E)$$

$$\frac{d\psi}{dt} = u\,, \quad \frac{dT}{dt} = 0\,, \quad \frac{d\gamma}{dt} = -1, \quad \frac{dG}{dt} = 0$$

Note that $C = (\gamma + t)\,\text{grad}\, G$, $d_C C^* = 0$ and the behaviour of φ ($\frac{d\varphi}{dt} = V^2$) is implied by the above equations.

The two scalar invariant T et G characterize the motion. The classification of motions will be done in accordance with the position of surfaces T = const. and G = const. (If T = s, they are the isentropic surfaces and the Lamb surfaces respectively).

a) General case: T et G are independent ($\text{grad}\, T \wedge \text{grad}\, G \neq 0$)

The T-surfaces (isentropic) and the G-surfaces (Lamb surfaces) are different.

b) If T is not constant and if $G = G(T)$, the T-surfaces are G-surfaces.

Replacing ψ par $\psi + \gamma G'_T$ we obtain:

$$V = \text{grad}\,\varphi + \psi\,\text{grad}\,T \qquad \text{(G)}$$

$$\frac{d\psi}{dt} = u - G'(T)\,, \quad \frac{dT}{dt} = 0$$

These equations characterize permanent oligotropic motions [6].

c) G is constant and T is not constant : it is a particular case of the above ($\frac{d\psi}{dt} = u$).

d) T is constant, G is not constant. The equations are written:

$$V = \text{grad}\,\varphi + \gamma\,\text{grad}\,G$$

$$\frac{d\gamma}{dt} = -1, \quad \frac{dG}{dt} = 0, \quad T = T_o$$

e) If T and G are constant we obtain the irrotational motions:

$$V = \text{grad}\,\varphi\,, \quad G = G_o\,, \quad T = T_o$$

Let us note: Any irrotational motion is not as above (Hamel's motions [26]). In fact, relation (12) shows that if the motion is irrotational and if one of the mappings T or G is constant, the other one is; but if neither G nor T is constant, G must depend on T and this case is the second one: an irrotational motion is oligotropic.

Irrotational motions do not correspond to particular values taken by first integrals or values convected by the stream.

We represent the flow of a complex fluid by fitting areas inside D_t. These areas do not overlap. They are convected by the stream. If every one has its own colour, it retains it along the motion.

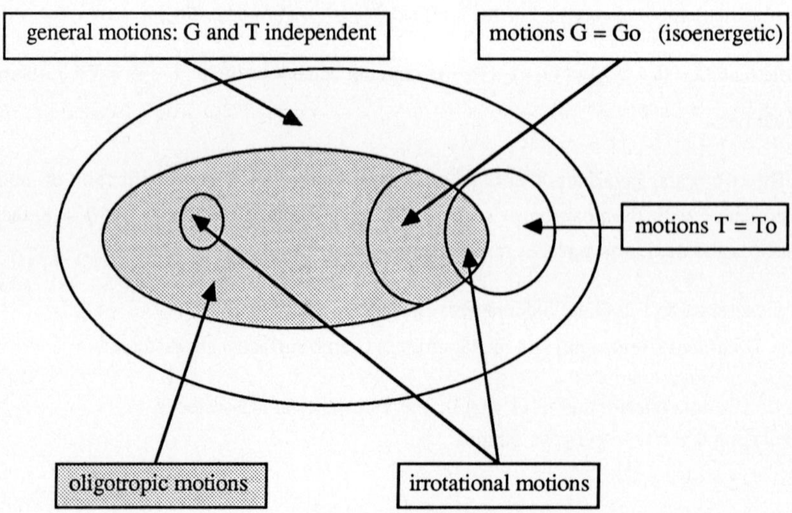

APPENDIX

Appendix 1

The motion of the medium consists in the t-dependent C^2 - diffeomorphism φ_t [5]:

$$X = (X^1, X^2, X^3) \in D_o \longrightarrow x = (x^1, x^2, x^3) \in D_t$$

We immediately deduce $\frac{dF}{dt} = \frac{\partial V}{\partial x} F$ and $\frac{dF^{-1}}{dt} = - F^{-1} \frac{\partial V}{\partial x}$. Let us write $T^*(D_t)$ for the cotangent fiber bundle of D_t and $T_x{}^*(D_t)$ the cotangent linear space to D_t at x; then,

$$x \in D_t \longrightarrow L(x,t) \in T_x{}^*(D_t)$$

represents a differential form fields on D_t.

Let us write $T^*(D_o)$ the cotangent fiber bundle of D_o and $T_X{}^*(D_o)$ the cotangent linear space to D_o at X. $\varphi_t{}^*$ is the mapping induced by φ_t for the form fields. The convective derivation d_c of a form field L is deduced by the diagram:

$$\begin{array}{ccc} L \in T^*(D_t) & \xrightarrow{\varphi_t{}^{-1*}} & L F \in T^*(D_o) \\ d_c \downarrow & & \downarrow \frac{d}{dt} \\ \frac{dL}{dt} + L \frac{\partial V}{\partial x} & \xleftarrow[\varphi_t{}^*]{} & \frac{dL}{dt} F + L \frac{\partial V}{\partial x} F \in T^*(D_o) \end{array} \tag{13}$$

$\frac{dL}{dt} + L \frac{\partial V}{\partial x}$ is the Lie derivative of L with respect to the velocity field V. (V is the infinitesimal transformation of the one-parameter group of transformation φ_t [27]).

Consequences :

Let b be a scalar field on D_t. D_t is assumed to be an euclidian space. We define two form fields by their values V* and (grad b)*. From (13) we deduce:

$$d_c (V^*) = \Gamma^* + \frac{\partial}{\partial x}\left(\frac{1}{2}V^2\right)$$

where Γ is the acceleration.

$$d_c (\text{grad } b)^* = \left(\text{grad} \frac{db}{dt}\right)^* \quad \text{and} \quad d_c (LF^{-1}) = \frac{dL}{dt} F^{-1}$$

and equation (2) gives:

$$d_c V^* = \text{grad}^*\left(\frac{1}{2}V^2 - h - \Omega\right) + \theta \ \text{grad}^* s \tag{14}$$

Appendix 2

Lemma: *Let* θ *and* s *be two differential scalars fields in* D_t.

There exist three scalar fields α, β, γ *such that:*

$$\theta \text{ grad}^* s = \alpha \text{ grad}^* \beta + \text{grad}^* \gamma \qquad \text{(a)}$$

is equivalent to :

there exists a function φ *such that :* $\beta = \varphi(\theta,s)$ (b)

- If $\text{grad } \theta \wedge \text{grad } s = 0 \Rightarrow \theta = \Phi'(s)$, $\theta \text{ grad } s = \text{grad } \Phi(s)$ and $\beta = 0$.
- If on an open set, $\text{grad } \theta \wedge \text{grad } s \neq 0$

(a) $\Rightarrow$ (b): $\text{grad } \theta \wedge \text{grad } s = \text{grad } \alpha \wedge \text{grad } \beta \Rightarrow$ grad θ , grad s , grad α , grad β are in the same plane and there exist φ, ψ, χ such that $\beta = \varphi(\theta,s)$ and $\alpha = \psi(\theta,s)$, $\gamma = \chi(\theta,s)$.

Conversely (b) $\Rightarrow$ (a): $\beta = \varphi(\theta,s)$ implies locally $\theta = h(\beta, s)$. But there exists $\gamma = H(\beta, s)$ such that $H'_s = \theta$. With $\alpha = - H'_\beta$, we obtain $\text{grad}^*\gamma = H'_s \text{ grad}^* s + H'_\beta \text{ grad}^*\beta$.

Appendix 3

Lemma: *Let* σ_o *be a differential scalar field in* D_o *, no constant and supposed to be independent of a given scalar field* δ_o *defined on* D_o *by* $\delta_o(X) = \delta[\varphi_t(X)]$.*(Such a field* σ_o *is arbitrary).*

For any form field $X \in D_o \longrightarrow \Xi_o^*(X) \in T_X^*(D_o)$ *there exist, at least locally, three scalar fields defined on* D_o *and denoted by* α_o , β_o , ω_o *such that*

$$\Xi_o^* = \text{grad}_o^* \alpha_o + \beta_o \text{ grad}_o^* \delta_o + \omega_o \text{ grad}_o^* \sigma_o$$

$\text{grad}_o^* \delta_o$ and $\text{grad}_o^* \sigma_o$ are two independent forms of D_o . The two surface families $\delta_o(X) = a_1$ and $\sigma_o(X) = b_1$ where a_1 and b_1 are two arbitrary constants, generate a family of curves denoted by $(\Gamma)_{a_1 b_1}$.

(S) being any transverse surface of this family, M_o denotes the current point of its intersection with $(\Gamma)_{a_1 b_1}$ and $\overline{M}_o\overline{M}$ the arc of $(\Gamma)_{a_1 b_1}$ with origin at M_o .

So, a differential function is defined by: $X \in D_o \longrightarrow \alpha_o(X) = \int_{\overline{M}_o\overline{M}} \Xi_o^* \, dX$ such that $d\delta_o = 0$, $d\sigma_o = 0$, $d\alpha_o = \Xi_o^* \, dX$.

Therefore, there exist, at least locally, two scalar field forms β_o and ω_o in D_o , such that

$$\Xi_o^* \, dX - d\alpha_o = \beta_o \, d\delta_o + \omega_o \, d\sigma_o$$

Theorem: *For any mapping* σ *of* D_t *, independent of* δ *and verifying* $\frac{d\sigma}{dt} = 0$, *for any form fields* Ξ^* *in* D_t *with a zero-convective derivation, there exist, at least locally, three mappings* α, β *and* ω *defined on* D_t *, verifying* $\frac{d\alpha}{dt} = 0$, $\frac{d\beta}{dt} = 0$, *and* $\frac{d\omega}{dt} = 0$, *such that* $\Xi^* = \text{grad}^* \alpha + \beta \text{ grad}^*\delta + \omega \text{ grad}^*\sigma$.

Let us write $\sigma(t,x) = \sigma_o[\varphi_t^{-1}(X)]$, $\alpha(t,x) = \alpha_o[\varphi_t^{-1}(X)]$, $\beta(t,x) = \beta_o[\varphi_t^{-1}(X)]$ and denote by Ξ^* the field of forms of D_t corresponding to Ξ_o^*.

σ, α, β, ω and δ have zero material derivatives, so forms and functions have a zero convective derivatives.

Consequently: With the representation of the form field C* and $\delta = T$, we can match φ and ψ in system (C) and we immediately obtain Clebsch's representation (or potential representation) for flows given by system (D).

Notice that the choice of C* is arbitrary and φ and ψ are not unique.

Proof of system (G):

System (G) is obtained when G is independent of T; in the other case, it is necessary to establish the proof for system (G). Only a small modification is needed to the previous lemma : if the form field $\Xi_o{}^*$ of D_o is given, there exist, at least locally, two scalar fields α_o and β_o such that:

$$\Xi_o{}^* = \text{grad}_o{}^* \alpha_o + \beta_o \, \text{grad}_o{}^* T_o$$

It is a representation of the same kind as established by Clesch [16].

In this way, let us consider a transverse curve (Γ) of T–surfaces and write $\alpha_o(X) = \int_{\overline{M}_o \overline{M}} \Xi_o{}^* \, dX$ for the real function defined on D_o by the circulation of $\Xi_o{}^*$ on an arbitrary plotted curve on the T–surface, joining X at the point X_o where this surface intersects (Γ).

This circulation is independent of the curve joining X_o at X and for $dT_o = 0$, $d\alpha_o = \Xi_o{}^* \, dX$.

So, there exists a Lagrangian multiplier $\beta_o(X)$ such that:

$$\Xi_o{}^* \, dX = d\alpha_o + \beta_o \, dT_o$$

and the Clebsch form is deduced.

REFERENCES

[1] GOUIN, H., Comptes rendus Acad. Sci. Paris, 305, II (1987) 833.

[2] BARRERE, M., and PRUD'HOMME, R., Equations fondamentales de l'aérothermochimie, Masson, Paris (1973)

[3] KORTEWEG, D.J., Archives Néerlandaises, 28, (1901) 1.

[4] SERRIN, J., Editor, New Perspectives in Thermodynamics, Springer, New York (1986).

[5] SERRIN, J., Encyclopedia of Physics VIII/1, Springer, Berlin (1959).

[6] CASAL, P., J. de mécanique, Paris, 5, n° 2 (1966) 149.

[7] GOUIN, H., J. de mécanique, Paris, 20, n° 2 (1981) 273.

[8] CASAL, P., and GOUIN, H., Sur les interfaces liquide-vapeur non-isothermes, J. de Mécanique Théorique et Appliquée, Paris, 7, n°6 (1988).

[9] CASAL, P., and GOUIN, H., A representation of liquid-vapour interfaces by using fluids of second grade, Annales de Physique, Paris, to appear (1988).

[10] SELIGER, R.L., and WHITHAM, G.B., Proc. Roy. Soc. of London, A, 305, (1968) 1.

[11] GOUIN, H., Research Notes in Mathematics, 46, Pitman, London (1981) 128.

[12] CASAL, P., and GOUIN, H., Comptes rendus Acad. Sci. Paris, 306, II (1988) 99.

[13] CUBISOL-GOUIN, F., Contribution à une étude thermomécanique des milieux continus, thesis, Université d'Aix-Marseille 3 (1987).

[14] TRUESDELL, C., and TOUPIN, R.A., The Classical Field Theories, Encyclopedia of Physics III/1, Springer, Berlin (1960).

[15] TRUESDELL, C., and NOLL, W., Encyclopedia of Physics III/3, Springer, Berlin (1965).

[16] LAMB, H., Hydrodynamics, Dover Publications, New York (1972).

[17] LANDAU, L., and LIFCHITZ, E., Fluid Mechanics, Mir, Moscow (1971).

[18] GOLDSWORTHY, F.A., J. Fluid Mech., 5, (1959) 164.

[19] STURTEVANT, B., and SLACHMUYDERS, E., Phys. Fluids, 7, (1964) 1201.

[20] MANDEL, J., Cours de mécanique des milieux continus, Gauthier-Villars, Paris (1966) 135.

[21] GOUIN, H., Mech. Res. Comm., 3, (1976) 151.

[22] MOREAU, J.J., Sur les intégrales premières de la dynamique d'un fluide parfait barotrope et le théorème de Helmholtz-Kelvin, Séminaire d'analyse convexe, 3, Montpellier (1977).

[23] CARTAN, E., Leçons sur les invariants intégraux, Hermann, Paris (1971).

[24] GOUIN, H., Contribution à une étude géométrique et variationnelle des milieux continus, Thesis, Université de Provence (1978).

[25] VILLAT, H., Leçons sur la théorie des tourbillons, Masson, Paris (1930).

[26] HAMEL, G., Sitzgsber. preus. Akad. Wiss. Phys. Math. Kl. (1937) 5.

[27] GERMAIN, P., Cours de mécanique des milieux continus, Masson, Paris (1973).

ON DIFFUSION IN TWO-PHASE SYSTEMS:
THE SHARP INTERFACE VERSUS THE TRANSITION LAYER

Morton E. Gurtin
Department of Mathematics
Carnegie-Mellon University
Pittsburgh, PA 15213

1. Introduction.

In [1986, 1988a, 1988b, 1988c] I began the development of a nonequilibrium thermodynamics of two-phase continua based on dynamical statements of the thermomechanical laws. The first three papers utilize Gibbs's [1878] notion of a sharp phase interface endowed with energy and entropy; [1988c] is based on ideas of van der Waals [1893], Korteweg [1901], and Cahn [1961, 1962] in which the interface is a smooth but thin transition layer, with interfacial energy accounted for by a free energy dependent on concentration gradients.

It is my purpose here to review and compare the results of these papers. The papers [1986, 1988a, 1988b] studied heat conduction, while [1988c] concerned isothermal diffusion; to have a common basis for discussion, I will present all of the theories within the framework of isothermal diffusion.

2. Single phase. Classical Diffusion.

To fix ideas, we begin by discussing the classical theory of diffusion for a binary mixture in a region Ω. The basic quantities, defined for all $x \in \Omega$ and all time t, are the **concentration** $c(x,t)$, the **mass flux** $\mathbf{h}(x,t)$, the **free energy** $\psi(x,t)$, and the **chemical potential** $\mu(x,t)$; the underlying laws, for any subregion P of Ω, are **balance of mass**

$$\{\textstyle\int_P c\}^{\bullet} = -\textstyle\int_{\partial P} \mathbf{h}\cdot\mathbf{n} \tag{2.1}$$ [1]

and the **second law**, which for isothermal diffusion has the form

$$\{\textstyle\int_P \psi\}^{\bullet} \le -\textstyle\int_{\partial P} \mu\mathbf{h}\cdot\mathbf{n} \quad , \tag{2.2}$$

where $\mathbf{n}$ is the outward unit normal to ∂P. Since P is arbitrary, these laws have

[1]We use the following notation: $\partial_z g$ denotes the derivative (usually partial) of g with respect to z; $g^{\bullet} = \partial_t g$ (t = time); for $\mathbf{m}$ a unit vector, $\partial_{\mathbf{m}} g$ is the surface gradient on the unit sphere. For a function $\phi(\omega)$ of a single scalar variable, $\phi'(\omega) = d\phi(\omega)/d\omega$.

local counterparts

$$\dot{c} = -\text{div}\,\mathbf{h}, \quad \dot{\psi} - \mu\dot{c} + \mathbf{h}\cdot\nabla\mu \leq 0 \quad . \tag{2.3}$$

The constitutive equations of the classical theory are

$$\psi = \psi(c), \quad \mu = \mu(c), \quad \mathbf{h} = -\mathbf{D}\nabla c \tag{2.4}$$

with $\mu'(c) > 0$ and $\mathbf{D}$, the **diffusivity**, a linear transformation; **compatibility**[1] with $(2.3)_2$ leads to the restrictions:

$$\mu(c) = \psi'(c), \quad \mathbf{D} \geq 0 \quad . \tag{2.5}$$

Balance of mass $(2.3)_1$ then reduces to the classical diffusion-equation

$$\dot{c} = \text{div}(\mathbf{D}\nabla c) \quad . \tag{2.6}$$

Alternatively, since $\mu(c)$ is invertible, we may take the chemical potential μ as independent variable; if we consider a given value μ_0 of μ, write $C = \mu_0 c'(\mu_0)$, $\mathbf{K} = \mu_0 C\mathbf{D}$, and $u = (\mu - \mu_0)/\mu_0$, and linearize (2.6) about μ_0, we arrive at

$$C\dot{u} = \text{div}(\mathbf{K}\nabla u) \quad . \tag{2.7}$$

3. Two phases. The sharp interface.[2]

3.1. Basic quantities.

We assume that: (i) Ω consists of two phases separated, at each t, by an **interface** $S(t)$, and we write $\Omega_i(t)$ for the subregion of the body occupied by phase i; (ii) $S(t)$ is a smoothly propagating surface with unit normal field $\mathbf{m}(x,t)$, the **orientation** of $S(t)$, chosen so that $\mathbf{m}(x,t)$ coincides with the outward unit normal to $\partial\Omega_1(t)$; (iii) $S(t)$ does not touch the boundary $\partial\Omega(t)$. We write $V(x,t)$ for the **normal velocity** of $S(t)$ in the direction $\mathbf{m}(x,t)$, $\mathbf{L}(x,t)$ for the **curvature tensor** on $S(t)$, and $H = \text{tr}\mathbf{L}$ for (twice) the **mean curvature**.

The behavior in **bulk** is described by the concentration $c(x,t)$, the mass flux $\mathbf{h}(x,t)$, the bulk free energy $\psi(x,t)$, and the chemical potential $\mu(x,t)$, with the chemical potential continuous across the interface, but the remaining bulk fields allowed to suffer jump discontinuities.

To describe the interface, we add three functions of t and $x \in S(t)$: the **interfacial free energy** $f(x,t)$, the **capillary stress** $\mathbf{C}(x,t)$, and the **interaction** $\boldsymbol{\pi}(x,t)$. The vector field $\boldsymbol{\pi}(x,t)$ characterizes the interaction between the interface and the bulk material, while $\mathbf{C}(x,t)$, a linear transformation of tangent vectors into (not necessarily tangent) vectors, represents microforces exerted across

[1]The requirement that all "smooth processes" related through (2.4) be consistent with $(2.3)_2$ (Coleman and Noll [1963]). In applying this procedure the assumption of appropriate supplies to insure mass balance (and later to insure balance of capillary forces) is left tacit.

[2]The results of this chapter are taken from [1988b].

∂S in response to the creation of new surface. We can always write $\mathbf{C}$ uniquely in the form $\mathbf{C} = \mathbf{C}_{tan} + \mathbf{m} \otimes \mathbf{c}$, where $\mathbf{C}_{tan}(x,t)$, the **tangential component**, is a linear transformation of tangent vectors into tangent vectors, while $\mathbf{c}(x,t)$, the **normal component**, is a tangent vector. We assume, without essential loss in generality, that $\mathbf{C}_{tan}(x,t)$ is symmetric (cf. [1988b], Remark 3.3).

3.2. Balance of capillary forces.

Let $E(t)$ be a sufficiently regular subsurface of $S(t)$, and let $\boldsymbol{\nu}(x,t)$, a vector field tangential to $S(t)$, be the outward unit normal to the boundary curve $\partial E(t)$. Then **balance of capillary forces** is the requirement that

$$\int_{\partial E} \mathbf{C}\boldsymbol{\nu} + \int_E \boldsymbol{\pi} = 0 \tag{3.1}$$

for all such E, or equivalently, that

$$\mathrm{div}_S \mathbf{C} + \boldsymbol{\pi} = 0 \tag{3.2}$$

with div_S the surface divergence on S. If we introduce the normal component $\pi = \boldsymbol{\pi}\cdot\mathbf{m}$ of the interaction, then the normal component of (3.2) has the form[1]

$$\mathbf{C}_{tan}\cdot\mathbf{L} + \mathrm{div}_S \mathbf{c} + \pi = 0 \quad ; \tag{3.3}$$

this relation is central to what follows.

We regard the normal velocity V as intrinsic in the sense that superficial forces supply power to the interface through the velocity $V\mathbf{m}$. This power is characterized by the identity:

$$\int_{\partial E} (V\mathbf{m})\cdot\mathbf{C}\boldsymbol{\nu} + \int_E V\mathbf{m}\cdot\boldsymbol{\pi} = -\int_E (V\mathbf{C}_{tan}\cdot\mathbf{L} + \mathbf{c}\cdot\mathbf{m}^{\Delta}) \quad . \tag{3.4}$$

The left side of (3.4) gives the total power expended on E, while the right side catalogs the manner in which this power is used: $-V\mathbf{C}_{tan}\cdot\mathbf{L}$ represents power expended in creating new surface; $-\mathbf{c}\cdot\mathbf{m}^{\Delta}$ represents power expended in changing the orientation of the interface. Here ϕ^{Δ} is the normal time derivative (the time derivative following the interface).

3.3. The second law.

Consider an arbitrary fixed subregion P of Ω with $\mathbf{n}$ the outward unit normal on ∂P, and let E be the portion of S that lies in P: $E(t) = P \cap S(t)$. We

---

[1]This normal component arises in previous theories as an Euler-Lagrange equation corresponding to the requirement that a global Gibbs function be stationary (Herring [1951], Cahn and Hoffman [1972, 1974], Fonseca [1988]). I believe that balance of capillary forces (3.1) is a basic law over and above the usual balance laws for forces and moments.

postulate that each such subregion P be consistent with **balance of mass** (2.1)[1] and the **second law** in the form

$$\{\int_P \psi + \int_E f\}^{\bullet} + \mathrm{outflow}(f,P) \leq -\int_{\partial P}\mu\mathbf{h}\cdot\mathbf{n} + \int_{\partial E}(V\mathbf{m})\cdot\mathbf{C}\boldsymbol{\nu} \quad ,$$

$$\mathrm{outflow}(f,P) := \int_{\partial E} fVp/(1-p^2)^{1/2}, \quad p = \mathbf{m}\cdot\mathbf{n} \quad , \tag{3.5}$$

where outflow(f,P) represents the rate at which interfacial free energy is carried out of P across ∂P due to the motion of the interface. Balance of mass and (3.5) lead to the **bulk relations** (2.3) and the **interface conditions**

$$[c]V = [\mathbf{h}]\cdot\mathbf{m} \quad ,$$

$$f^{\Delta} + \mathbf{c}\cdot\mathbf{m}^{\Delta} + V\{-fH + \mathbf{C}_{tan}\cdot\mathbf{L} - [G] + \pi\} \leq 0 \quad , \tag{3.6}$$

where $[\varphi]$ denotes the jump in a function φ across the interface (phase 2 minus phase 1), while G is the **Gibbs function**

$$G := \psi - \mu c \quad . \tag{3.7}$$

3.4. Constitutive equations. Thermodynamic restrictions.

We consider bulk constitutive equations in the form (2.4) subject to the restrictions (2.5), but we allow the two phases to exhibit different response:

$$\psi = \psi_i(c), \quad \mu_i(c) = \psi_i'(c), \quad \mathbf{h} = -\mathbf{D}_i\nabla c \quad . \tag{3.8}$$

As before, we assume that $\mu_i'(c) > 0$; this allows us to use μ as independent variable, in which case (3.7) and (3.8) yield

$$G = G_i(\mu), \quad c = c_i(\mu) = -G_i'(\mu) \quad . \tag{3.9}$$

We consider a fairly general constitutive theory for the interface. The free energy f is allowed to depend on the chemical potential μ, and - to have a theory of sufficient generality to model crystal growth - also on the orientation $\mathbf{m}$ and the normal velocity V; in addition, constitutive equations are given for the tangential and normal components, $\mathbf{C}_{tan}$ and $\mathbf{c}$, of the capillary stress and for the normal component π of the interaction:

$$f = f(\mu,\mathbf{m},V), \qquad \pi = \pi(\mu,\mathbf{m},V) \quad ,$$

$$\mathbf{C}_{tan} = \mathbf{C}_{tan}(\mu,\mathbf{m},V), \qquad \mathbf{c} = \mathbf{c}(\mu,\mathbf{m},V) \quad .$$

Compatibility with $(3.6)_2$ then leads to the following list of constitutive restrictions:

(i) the free energy and the normal and tangential components of the capillary stress are independent of the chemical potential μ and the normal velocity V and

---

[1]To simplify the presentation, we neglect interfacial mass.

$$\mathbf{c}(\mathbf{m}) = -\partial_{\mathbf{m}} f(\mathbf{m}) \quad ,$$

$$\mathbf{C}_{\tan}(\mathbf{m}) \text{ is a surface tension of amount } f(\mathbf{m}) \quad ; \tag{3.10}$$

(ii) the interaction has the form

$$\pi = G_2(\mu) - G_1(\mu) - \beta(\mu,\mathbf{m},V)V, \quad \beta > 0 \quad . \tag{3.11}$$

3.5. Approximate free-boundary problems.

We assume now that the bulk Gibbs functions $G = G_i(\mu)$ coincide at a unique value $\mu = \mu_M$, called the **transition potential**. In classical theories without interfacial structure changes in phase occur at the transition potential. Within our theory a flat and stationary interface has potential μ_M, but a curved or a moving interface need not; in fact, the relation

$$u = f(\mathbf{m})H + \partial_{\mathbf{m}}\partial_{\mathbf{m}} f(\mathbf{m})\cdot\mathbf{L} - \beta(\mathbf{m})V \quad , \tag{3.12}$$

for the (dimensionless) potential $u = (\mu - \mu_M)/\mu_M$ as a function of V, the curvature tensor $\mathbf{L}$, and the corresponding mean curvature H, follows from capillary balance as an approximation appropriate to an interface with small interfacial densities and weak dependence on V. Here $\beta(\mathbf{m}) = \beta(\mu_M,\mathbf{m},0)$, we have assumed that $c_2(\mu_M) > c_1(\mu_M)$, and we have chosen an appropriate scaling.

The diffusion equation (2.7) for each phase (in bulk) combined with (3.12) and a similar approximation for balance of mass $(3.6)_1$ lead to the partial differential equations and free-boundary conditions:[1]

$$\begin{aligned} &C_i u^{\bullet} = -\mathrm{div}\,\mathbf{h}, \quad \mathbf{h} = -\mathbf{K}_i \nabla u \quad \text{in bulk,} \\ &\left.\begin{aligned} &u = \mathbf{B}(\mathbf{m})\cdot\mathbf{L} - \beta(\mathbf{m})V \\ &V = [\mathbf{h}]\cdot\mathbf{m} \end{aligned}\right\} \quad \text{on the interface;}^2 \end{aligned} \tag{3.13}$$

$\mathbf{B}(\mathbf{m})\cdot\mathbf{L}$ represents the first two terms on the right side of (3.12). For quasi-static diffusion $(C_i = 0)$ in an isotropic material with $\beta = 0$, (3.13) reduces to the Mullins-Sekerka [1963, 1964] system

$$\begin{aligned} &\Delta u = 0, \quad \mathbf{h} = -k_i \nabla u \quad \text{in bulk} \\ &\left.\begin{aligned} &u = fH \\ &V = [\mathbf{h}]\cdot\mathbf{m} \end{aligned}\right\} \quad \text{on the interface} \quad , \end{aligned} \tag{3.14}$$

---

[1]The phrase "in bulk" is shorthand for all $x \in \Omega_i(t)$ and all t; an analogous interpretation applies to "on the interface".

[2]The condition $V = [\mathbf{h}]\cdot\mathbf{m}$ is a classical Stefan condition. Free-boundary conditions of the form $u = -\beta(\mathbf{m})V$ were introduced by Frank [1958]; $u = fH$ (f=constant) was introduced by Mullins and Sekerka [1963, 1964]; $u = fh - \beta V$ was used by Voronkov [1964].

with $f, k_i > 0$ scalar constants and Δ the Laplacian.

Global growth conditions follow for the general system (3.13) in a bounded domain under various boundary conditions. In particular, for $C_1 = C_2 = C$ and $\partial\Omega$ held at constant potential $u = U$,

$$\{\int_S f(\mathbf{m}) + U\mathrm{vol}(\Omega_1) + (C/2)\int_\Omega (u-U)^2\}^\bullet \le 0 \quad . \tag{3.15}$$

4. Gradient theory of phase transitions.[1]

4.1. Basic quantities. The second law.

The theory discussed in Section 3 models the transition between phases as a **sharp** interface; that is, as a surface of zero thickness, endowed with a thermodynamical structure separate from that of the bulk material. Alternatively, Cahn [1961, 1962][2] has proposed a dynamical theory of phase transitions in which the phase interface is a **smooth** but thin **transition layer**; within this layer concentration gradients are large, and interfacial energy is accounted for by allowing the free energy to depend on concentration gradients. This theory was framed within the context of continuum thermodynamics in [1988c]; we now sketch the main ideas of [1988c].

The primitive quantities, defined for all $x \in \Omega$ and all t, are the concentration $c(x,t)$, the mass flux $\mathbf{h}(x,t)$, the free energy $\psi(x,t)$, and the chemical potential $\mu(x,t)$, augmented by two new quantities: the **capillarity potential** $\lambda(x,t)$ and the **capillarity flux** $\zeta(x,t)$. Roughly speaking, the capillarity potential characterizes the kinetics of the transition layer, while the capillarity flux represents microforces involved in the motion of this layer.

The basic physical laws, for any subregion P of Ω, are **balance of mass** (2.1) and the **second law**

$$\{\int_P \psi\}^\bullet \le -\int_{\partial P} \mu\mathbf{h}\cdot\mathbf{n} + \int_{\partial P} \lambda^\bullet \boldsymbol{\zeta}\cdot\mathbf{n} \tag{4.1}$$

with $\mathbf{n}$ the outward unit normal to ∂P. The last term represents mechanical power expended on the transition zone and is the analog of the last term in (3.5) of the sharp theory. Since P is arbitrary, (2.1) and (4.1) have the local forms

$$c^\bullet = -\mathrm{div}\,\mathbf{h}, \qquad \psi^\bullet - \mu c^\bullet - \mathrm{div}(\lambda^\bullet \zeta) + \mathbf{h}\cdot\nabla\mu \le 0 \ . \tag{4.2}$$

4.2. Constitutive equations. Thermodynamic restrictions.

Essential to the theory are constitutive equations which include a dependence on higher gradients:

---

[1]The results of this section are taken from [1988c].

[2]See also Cahn and Hilliard [1971]. The use of higher gradients to model capillarity is due to van der Waals [1893] and Korteweg [1901].

$$\psi = \psi(c,\nabla c,\nabla^2 c,\nabla^3 c,\dots), \qquad \mu = \mu(c,\nabla c,\nabla^2 c,\nabla^3 c,\dots) \ ,$$

$$\mathbf{h} = \mathbf{h}(c,\nabla c,\nabla^2 c,\nabla^3 c,\dots), \qquad \lambda = \lambda(c,\nabla c,\nabla^2 c,\nabla^3 c,\dots) \ , \tag{4.3}$$

$$\zeta = \zeta(c,\nabla c,\nabla^2 c,\nabla^3 c,\dots) \ .$$

It is useful to define an additional field, the **reduced capillarity vector** $\boldsymbol{\xi} = \boldsymbol{\xi}(c,\nabla c,\nabla^2 c,\nabla^3 c,\dots)$, through

$$\boldsymbol{\xi} = (\partial_c \lambda)\boldsymbol{\zeta} \ , \tag{4.4}$$

and to write

$$\mathbf{g} = \nabla c, \qquad \mathbf{G} = \nabla^2 c \ .$$

Under fairly weak assumptions, compatibility with $(4.2)_2$ requires that the free energy can depend at most on c and $\mathbf{g}$, the chemical potential at most on c, $\mathbf{g}$, and $\mathbf{G}$. In fact,

$$\psi = \psi(c,\mathbf{g}), \qquad \mu = \mu(c,\mathbf{g},\mathbf{G}), \qquad \lambda = \lambda(c), \qquad = \boldsymbol{\xi}(c,\mathbf{g}) \tag{4.5}$$

with

$$\mu = \partial_c\psi - \operatorname{div}\boldsymbol{\xi}, \qquad \boldsymbol{\xi} = \partial_{\mathbf{g}}\psi, \qquad \mathbf{h}\cdot\nabla\mu \le 0 \ . \tag{4.6}$$

Further, since $\lambda = \lambda(c)$, (4.4) yields (cf. (4.1))

$$\int_{\partial P} \dot{\lambda}\,\boldsymbol{\zeta}\cdot\mathbf{n} = \int_{\partial P} \dot{c}\,\boldsymbol{\xi}\cdot\mathbf{n}$$

for any subregion P, so that, without loss in generality, we may consider c as capillarity potential and $\boldsymbol{\xi}$ as capillarity flux:

$$\lambda = c, \qquad \boldsymbol{\zeta} = \boldsymbol{\xi} \ . \tag{4.7}$$

Also studied in [1988c] is a **quasi-linear theory** in which the chemical potential, mass flux, and reduced capillarity vector are affine functions of the gradients of c. In this case compatibility with $(4.2)_2$ yields

$$\begin{aligned} \psi &= \psi_0(c) + \tfrac{1}{2}\,\mathbf{g}\cdot\mathbf{A}\mathbf{g} \ , \\ \mu &= \psi_0'(c) - \mathbf{A}\cdot\mathbf{G} \ , \\ \boldsymbol{\xi} &= \mathbf{A}\mathbf{g} \ , \\ \mathbf{h} &= -\mathbf{K}(c)\nabla\mu \ , \end{aligned} \tag{4.8}$$

where $\mathbf{A}$ and $\mathbf{K}(c)$ are second-order tensors with $\mathbf{A}$ symmetric, $\mathbf{A} \ge 0$, $\mathbf{A} \ne 0$, $\mathbf{K}(c) \ge 0$. Here $\psi_0(c)$ represents the **coarse-grain free-energy** (the free energy at constant concentration).

The constitutive equations (4.8), when combined with $(4.2)_1$, yield a single partial differential equation for the concentration,

$$\dot{c} = \operatorname{div}\{\mathbf{K}(c)\nabla[\psi_0'(c) - \mathbf{A}\cdot\nabla^2 c]\} \ , \tag{4.9}$$

which, for an isotropic material with $\mathbf{K}$ constant, reduces to the **Cahn-Hilliard**

equation (Cahn [1961, 1962]; Cahn and Hilliard [1971])

$$c^{\bullet} = k\Delta[\psi_0'(c) - a\Delta c] \quad . \tag{4.10}$$

5. Comparison of the smooth and sharp theories.

5.1. The formal result.

Since the "smooth" and "sharp" theories discussed above pruportedly model the same physical phenomena, a comparison seems relevant. Balance of mass is essentially the same for both theories; what differs is the second law. We now give a formal argument which at least indicates that for small capillarity the two versions of the second law are consistent.

The second law for the two theories has the forms:

sharp theory

$$\{\int_P \psi + \int_E f\}^{\bullet} + \text{outflow}(f,P) \leq -\int_{\partial P} \mu \mathbf{h}\cdot\mathbf{n} + \int_{\partial E} Vc\cdot\boldsymbol{\nu} \quad ; \tag{5.1}$$

smooth theory

$$\{\int_P \psi\}^{\bullet} \leq -\int_{\partial P} \mu \mathbf{h}\cdot\mathbf{n} + \int_{\partial P} c^{\bullet}\boldsymbol{\xi}\cdot\mathbf{n} \quad , \tag{5.2}$$

For the purpose of comparison we will consider the smooth theory within the quasi-linear framework (4.8) with $\mathbf{K} > 0$ constant and $\mathbf{A} > 0$.

In the smooth theory the two phases are described by the coarse-grain free-energy $\psi_0(c)$, which is nonconvex as shown in Figure 1. The minimum values of $\psi_0(c)$ at c_1 and c_2 are, for convenience, $\psi_0(c_1) = \psi_0(c_2) = 0$. The portion of $(0,1)$ on which ψ_0 is locally convex has two connected components, the **two phases** C_1 and C_2 (labeled so that $c_i \in C_i$). C_1 and C_2 are separated by the **spinodal** S, consisting of all c with $\psi''(c) < 0$.

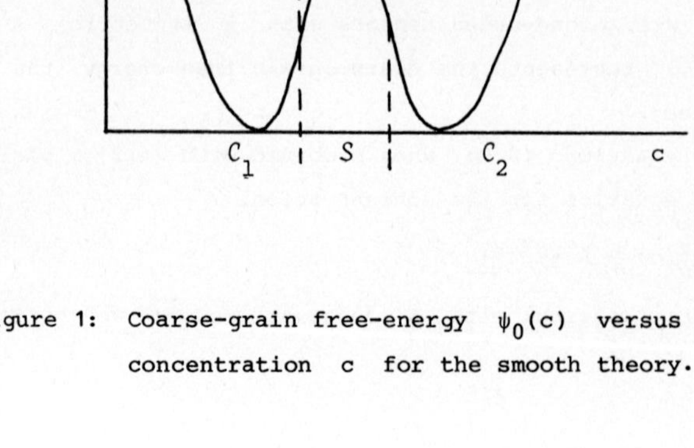

Figure 1: Coarse-grain free-energy $\psi_0(c)$ versus concentration c for the smooth theory.

The term $\frac{1}{2}\nabla c\bullet \mathbf{A}\nabla c$ in the constitutive equation for the free energy models capillarity in the transition region between phases. In fact, results of Cahn and Hilliard [1958] and Carr, Gurtin, and Slemrod [1984] for the one-dimensional statical theory indicate that, for a concentration field c which varies only in the direction $\mathbf{m}$,

$$\sigma(\mathbf{m}) := \beta(\tfrac{1}{2}\,\mathbf{m}\bullet\mathbf{Am})^{1/2}, \qquad \beta = 2\int_{C_1}^{C_2}\psi_0(c)^{1/2}\,dc \tag{5.3}$$

is an appropriate choice for the interfacial free-energy per unit area. Guided by $(3.10)_1$, we define

$$\mathbf{t}(\mathbf{m}) := -\partial_{\mathbf{m}}\sigma(\mathbf{m}) = -\frac{1}{2}\beta^2\sigma(\mathbf{m})^{-1}\mathbf{P}(\mathbf{m})\mathbf{Am}\ , \tag{5.4}$$

where $\mathbf{P}(\mathbf{m})$ is the perpendicular projection onto the plane perpendicular to $\mathbf{m}$.

To avoid complications associated with $\partial\Omega$, we take $\Omega = \mathbf{R}^3$ and consider a solution $c(x,t)$ of (4.9) defined for all $x \in \mathbf{R}^3$ and all t in some time interval T. We write ℓ for the minimum diameter of the bounded phase region at some given time $t_0 \in T$ (for example, the initial time) and we introduce dimensionless quantities

$$\begin{gathered} x^* = x/\ell, \quad t^* = t\beta^2|\mathbf{K}|/\ell^2, \quad \psi_0^* = \psi_0/\beta^2, \quad \mathbf{A}^* = \mathbf{A}/\ell^2\beta^2,\ \mathbf{K}^* = \mathbf{K}/|\mathbf{K}|\ , \\ \sigma^*(\mathbf{m}) = \sigma(\mathbf{m})/\ell\beta^2 = (\tfrac{1}{2}\,\mathbf{m}\bullet\mathbf{A}^*\mathbf{m})^{1/2},\ t^*(\mathbf{m}) = -\frac{1}{2}\sigma^*(\mathbf{m})^{-1}\mathbf{P}(\mathbf{m})\mathbf{A}^*\mathbf{m} = -\partial_{\mathbf{m}}\sigma^*(\mathbf{m}) \end{gathered} \tag{5.5}$$

and dimensionless constitutive variables

$$\psi^* = \psi/\beta^2, \quad \mu^* = \mu/\beta^2, \quad \boldsymbol{\xi}^* = \boldsymbol{\xi}/\ell\beta^2, \quad \mathbf{h}^* = \mathbf{h}\ell/|\mathbf{K}|\beta^2\ . \tag{5.6}$$

We assume that $\mathbf{A}^*$ is small in the sense that $\mathbf{A}^* = \delta^2\mathbf{B}$ with $\delta > 0$ a small parameter and $\mathbf{B}$ a fixed tensor; then, trivially,

$$\mathbf{A}^* = O(\delta^2), \qquad \sigma^*(\mathbf{m}) = O(\delta)\ . \tag{5.7}$$

For convenience, we drop the superscript $*$ and note, for future use, that the constitutive equations (4.8) and the differential equation (4.9) remain unchanged.

We are now in a position to establish[1] the **formal consistency of the two forms of the second law:** for small δ the second law (5.2) for the smooth theory is asymptotic (cf. (5.28)) to the second law (5.1) for the sharp theory provided we make the following identifications:

---

[1]This result, presented at the October 1987 meeting of the Society for Natural Philosophy, but not published, led me to conjecture, at that meeting, that a suitable version of the Cahn-Hilliard equation (4.10) regularizes the Mullins-Sekerka system (3.14). Independently, Pego [1988] gave a formal asymptotic analysis indicating that this is in fact the case. The asymptotic equivalence of the two models leads me to believe that the anisotropic equation (4.9) might yield instabilities of the type studied via the Mullins-Sekerka system ([1963, 1964]). Here it would seem appropriate to consider: (i) boundary conditions of the form $c = c_b$ and $\mu = \psi_0'(c_b)$ on $\partial\Omega$ with c_b metastable in the sense that $c_b \in \mathcal{C}_1$, $c_b > C_1$; (ii) initial conditions with $c(x,0) > C_2$ in a small set and close to c_b elsewhere.

bulk	
sharp theory	smooth theory
$\psi_i(c)$	$\psi_0(c),\ c \in \mathcal{C}_i$
$\mu_i(c)$	$\psi_0'(c),\ c \in \mathcal{C}_i$

interface	
sharp theory	smooth theory
$f(\mathbf{m})$	$\sigma(\mathbf{m})$
$\mathbf{c}(\mathbf{m})$	$\mathbf{t}(\mathbf{m})$

5.2. Formal analysis.

We assume that both phases are present at each t, so that all concentrations c_s in the spinodal $\mathcal{S}$ are represented. Choose a particular $c_s \in \mathcal{S}$ and let

$$S(t) := \{x \in \Omega: c(x,t) = c_s\} \quad . \tag{5.8}$$

We assume that: (i) $S(t)$ is a closed oriented surface with unit normal field $\mathbf{m}(y,t)$; (ii) there is a number $h(\delta) > 0$ such that

$$\mathcal{T}_\delta(t) := \{x \in \mathbf{R}^3: x = y + \lambda\mathbf{m}(y,t),\ y \in S(t),\ -h(\delta) < \lambda < h(\delta)\}$$

contains the spinodal region $\{x \in \mathbf{R}^3: c(x,t) \in \mathcal{S}\}$; (iii) $h(\delta)$ approaches zero with δ, but at a slightly slower rate; precisely,

$$h(\delta) \to 0, \quad \delta^{-1}h(\delta) \to \infty, \quad \delta^{-1}h^2(\delta) \to 0 \quad \text{as} \quad \delta \to 0 \tag{5.9}$$

(e.g., $h(\delta) = \delta\ell n\delta$); (iv) the mapping $(\lambda,y) \to x = y + \lambda\mathbf{m}(y,t)$ is a bijection of $[-h(\delta),h(\delta)] \times S(t)$ onto $\mathcal{T}_\delta(t)$. We will refer to $\mathcal{T}_\delta(t)$ as the **transition layer**; there $c(x,t)$ may be considered a function $c(\lambda,y,t)$ of t and (λ,y). We assume that: (v) the transition layer separates phases,

$$c(-h(\delta),y,t) \in \mathcal{C}_1, \qquad c(h(\delta),y,t) \in \mathcal{C}_2 \quad ; \tag{5.10}$$

(vi) outside the layer $c(x,t)$ experiences only moderate variations:

$$\text{outside of } \mathcal{T}_\delta(t)\text{: } c \text{ is bounded away from zero and one; } \dot{c},\ \nabla c,\ \nabla^2 c,\ \nabla^3 c = O(1) \quad . \tag{5.11}$$

We expect that $c(\lambda,y,t)$ varies little with y, but that variations with λ are large, and we scale λ by defining

$$\theta = \lambda/\sigma(\mathbf{m}), \qquad \mathbf{m} = \mathbf{m}(y,t) \quad . \tag{5.12}$$

Then to each $x \in \mathcal{T}_\delta(t)$ there correspond coordinates $\theta = \Theta(x,t)$, $y = \mathbf{Y}(x,t)$. We will develop an asymptotic analysis relevant to (θ,y) as spatial variables. Let c_δ and μ_δ denote c and μ with the dependence on δ explicit. We consider formal expansions

$$c_\delta(\theta,y,t) = c_0(\theta,y,t) + O(\delta) \quad ,$$
$$\mu_\delta(\theta,y,t) = \mu_0(\theta,y,t) + O(\delta) \qquad (5.13)$$

with θ and y derivatives of the $O(\delta)$ terms also of $O(\delta)$. Since θ varies between $\pm h(\delta)/\sigma(\mathbf{m})$ with $\sigma(\mathbf{m}) = O(\delta)$, it is clear from (5.9) that c_0 and μ_0 should be defined for $-\infty < \theta < \infty$. For convenience, we henceforth write $c = c_\delta$, $\mu = \mu_\delta$.

Given a function $\varphi(x,t)$, we write $\nabla\varphi$ for the gradient with respect to x, $\varphi^\bullet$ for the time derivative holding x fixed, and φ' for the derivative with respect to θ holding y and t fixed. Then, since

$$\theta(x,t)\sigma(\mathbf{m})\mathbf{m} = -\mathbf{Y}(x,t) + x, \qquad \mathbf{m} = \mathbf{m}(\mathbf{Y}(x,t),t) \quad , \qquad (5.14)$$

it follows that

$$\nabla\theta = \sigma(\mathbf{m})^{-1}\mathbf{m} + O(1), \quad \nabla^2\theta = O(\delta^{-1}), \quad \nabla\mathbf{Y}, \nabla^2\mathbf{Y} = O(1) \quad , \qquad (5.15)$$

and, for φ equal to c or μ,

$$\nabla\varphi = \sigma(\mathbf{m})^{-1}\varphi_0'\mathbf{m} + O(1), \quad \nabla^2\varphi = \sigma(\mathbf{m})^{-2}\varphi_0''\mathbf{m}\otimes\mathbf{m} + O(\delta^{-1}) \quad . \qquad (5.16)$$

Let $V(y,t)$ denote the normal velocity of $S(t)$ at $y \in S(t)$ in the direction $\mathbf{m}(y,t)$. We assume that in the current time scale $V, \mathbf{m}^\bullet = O(1)$. In $\mathcal{T}_\delta$, the level curves of c will have normal velocity $\pm c^\bullet/|\nabla c|$. Since $S(t)$ is such a curve, this assumption is consistent with the assumption that

$c_0^\bullet = O(\delta^{-1})$, and the $O(\delta)$ term in
(5.13)$_1$ has a time derivative of $O(1)$. (5.17)

Further, $V = \mathbf{Y}^\bullet \cdot \mathbf{m}$ and, by (5.14),

$$\theta^\bullet(x,t) = -\sigma(\mathbf{m})^{-1}V(y,t) + O(1) \quad . \qquad (5.18)$$

Let $F(c) = d\psi_0(c)/dc$. By (4.8)$_2$, (4.9), (5.16), and (5.17),

$$[F(c_0) - 2c_0'']'' = 0, \quad \mu_0 = F(c_0) - 2c_0'' \quad . \qquad (5.19)$$

Consistency of (5.13) with (5.10) and (5.11) implies, by virtue of (5.15), that c_0 is bounded away from zero and one, and

$$c_0(-\infty,y,t) \in \mathcal{C}_1, \quad c_0(+\infty,y,t) \in \mathcal{C}_2, \quad c_0'(\pm\infty,y,t) = c_0''(\pm\infty,y,t) = 0 \quad . \qquad (5.20)$$

By (5.19) and (5.20), μ_0 is independent of θ, while (5.8) yields $c_0(0) = c_s$. Thus, in view of the properties of ψ_0, it is clear from the phase portrait of $2c_0'' = F(c_0) - \mu_0$ that (5.19)$_2$ and (5.20) have a unique solution, and this solution has the properties:

$$c_0(\theta,y,t) = c_0(\theta) \text{ is independent of } y \text{ and } t \quad ,$$
$$\mu_0 = 0, \quad c_0(-\infty) = C_1, \quad c_0(\infty) = C_2, \quad c_0' > 0 \qquad (5.21)$$

(cf. Cahn and Hilliard [1958]). Thus, by (5.17) and (5.18),

$$c^{\bullet}(x,t) = -\sigma(\mathbf{m})^{-1}c_0'(\theta)V(y,t) + O(1) \quad . \tag{5.22}$$

Moreover, since $2c_0'' = F(c_0)$, $(5.3)_2$, (5.5), and the fact that ψ_0 is actually ψ_0^* yield

$$\psi_0(c_0) = (c_0')^2, \quad \int_{-\infty}^{\infty}(c_0')^2 d\theta = \int_{C_1}^{C_2}\psi_0(a)^{1/2}\, da = \frac{1}{2} \quad . \tag{5.22}$$

Next, since $\psi_0(c) = \frac{1}{2}\psi_0''(C_i)(c-C_i)^2 + O((c-C_i)^3)$ near $c = C_i$, (5.21) and $(5.22)_1$ imply that $c_0'(\theta) = O(e^{-\tau|\theta|})$ as $|\theta| \to \infty$ $(\tau > 0)$; by (5.9), (5.12), and (5.22), this yields

$$\sigma(\mathbf{m})^{-1}\int_{-h(\delta)}^{h(\delta)}(c_0')^2 d\lambda = \int_{-\infty}^{\infty}(c_0')^2 d\theta + o(1) = \frac{1}{2} + o(1) \quad . \tag{5.23}$$

Writing $\psi(c,\nabla c) = \psi_0(c) + \frac{1}{2}\nabla c\bullet\mathbf{A}\nabla c$ and using (5.16) and (5.21)-(5.23), we have the estimates

$$\psi(c,\nabla c) = 2(c_0')^2 + O(\delta), \quad \mu = O(\delta), \quad \nabla\mu = O(1) \quad \text{in } T_\delta \;,$$

$$c^{\bullet}\nabla c = -\sigma(\mathbf{m})^{-2}(c_0')^2 V\mathbf{m} + O(\delta^{-1}) \quad \text{in } T_\delta \;, \tag{5.24}$$

$$\sigma(\mathbf{m})^{-1}\int_{-h(\delta)}^{h(\delta)}\psi(c,\nabla c)d\lambda = 1 + o(1), \quad \sigma(\mathbf{m})\int_{-h(\delta)}^{h(\delta)} c^{\bullet}\nabla c\, d\lambda = -\frac{1}{2}\mathbf{m}V + o(1) \quad .$$

Consider now a given region P. Let $\mathbf{n}$ denote the outward unit normal to ∂P, and let $E(t) = P \cap S(t)$. We assume that $S(t)$ is nowhere tangential to ∂P, and that the intersection of $S(t)$ with ∂P is a smooth boundary curve $\partial E(t)$. We let $\boldsymbol{\nu}(y,t)$ denote the outward unit normal to $\partial E(t)$, so that $\boldsymbol{\nu}(y,t)$ is a vector field on $\partial E(t)$ tangent to $S(t)$.

Our next step is to estimate the integral of $\psi(c,\nabla c)$ over $P \cap T_\delta$ and the integrals of $c^{\bullet}\boldsymbol{\xi}\bullet\mathbf{n} = \mathbf{n}\bullet(\mathbf{A}\nabla c)c^{\bullet}$ and $\mu\mathbf{h}\bullet\mathbf{n} = -\mu\mathbf{n}\bullet\mathbf{K}\nabla\mu$ over $\partial P \cap T_\delta$. Since $\mathbf{n}(\lambda,y,t) = \mathbf{n}(0,y,t) + O(h(\delta))$, (5.9) and (5.24) yield

$$\int_{P\cap T_\delta}\psi(c,\nabla c) = \int_E\Big(\int_{-h(\delta)}^{h(\delta)}\psi(c,\nabla c)d\lambda\Big)dy + o(\delta) = \int_E \sigma(\mathbf{m}) + o(\delta) \quad ,$$

$$\int_{\partial P\cap T_\delta} c^{\bullet}\mathbf{n}\otimes\nabla c = \int_{\partial E}(\boldsymbol{\nu}\bullet\mathbf{n})^{-1}\mathbf{n}\otimes\Big(\int_{-h(\delta)}^{h(\delta)} c^{\bullet}\nabla c\, d\lambda\Big)dy + o(\delta^{-1})$$

$$= -\frac{1}{2}\int_{\partial E}V\sigma(\mathbf{m})^{-1}(\boldsymbol{\nu}\bullet\mathbf{n})^{-1}\mathbf{n}\otimes\mathbf{m} + o(\delta^{-1}) \quad ,$$

$$\int_{\partial P\cap T_\delta} c^{\bullet}\boldsymbol{\xi}\bullet\mathbf{n} = -\frac{1}{2}\int_{\partial E}V\sigma(\mathbf{m})^{-1}(\boldsymbol{\nu}\bullet\mathbf{n})^{-1}\mathbf{A}\mathbf{n}\bullet\mathbf{m} + o(\delta) \quad ,$$

$$\int_{\partial P\cap T_\delta}\mu\mathbf{h}\bullet\mathbf{n} = o(\delta) \quad .$$

Let $P_\delta = P_\delta(t)$ and $(\partial P)_\delta = (\partial P)_\delta(t)$ denote the portions of P and ∂P that lie <u>outside</u> of $T_\delta(t)$, and let

$$P_i = P_i(\delta,t) := \{x \in (\partial P)_\delta : c(x,t) \in C_i\} \quad ,$$

$$(\partial P)_i = (\partial P)_i(\delta,t) := \{x \in (\partial P)_\delta : c(x,t) \in C_i\} \quad ;$$

P_i and $(\partial P)_i$ denote the phase i portions of P and ∂P that lie outside of the transition layer. Further, for $c \in C_i$, let

$$\psi_i(c) = \psi_0(c), \quad \mu_i = \psi_i'(c), \quad \mathbf{h}_i = -\mathbf{K}\nabla\mu_i$$

$(\psi_i' = d\psi_i/dc)$. Then (4.8), (5.7), and (5.11) yield

$$\int_{P_\delta}\psi(c,\nabla c) = \Sigma \int_{P_i}\psi_i(c) + o(\delta),$$

$$\int_{(\partial P)_\delta} c^{\bullet}\xi\cdot\mathbf{n} = o(\delta), \quad \int_{(\partial P)_\delta}\mu\mathbf{h}\cdot\mathbf{n} = \Sigma\int_{(\partial P)_\delta}\mu_i\mathbf{h}_i\cdot\mathbf{n} + o(\delta)$$

where the sums are over $i = 1,2$. Thus, by (5.25),

$$\begin{aligned}
&\int_P\psi(c,\nabla c) = \Sigma\int_{P_i}\psi_i(c) + \int_E \sigma(\mathbf{m}) + o(\delta) \quad , \\
&\int_{\partial P} c^{\bullet}\xi\cdot\mathbf{n} = -\frac{1}{2}\int_{\partial E} V\sigma(\mathbf{m})^{-1}(\nu\cdot\mathbf{n})^{-1}\mathbf{A}\mathbf{n}\cdot\mathbf{m} + o(\delta) \quad , \qquad (5.26)\\
&\int_{\partial P}\mu\mathbf{h}\cdot\mathbf{n} = \Sigma\int_{(\partial P)_i}\mu_i\mathbf{h}_i\cdot\mathbf{n} = o(\delta) \quad .
\end{aligned}$$

Finally, by $(3.5)_2$, (5.4), and the identities $\mathbf{n} = (\mathbf{n}\cdot\mathbf{m})\mathbf{m} + (\mathbf{n}\cdot\nu)\nu$ and $2\sigma(\mathbf{m})^2 = \mathbf{m}\cdot\mathbf{A}\mathbf{m}$, $(5.26)_2$ takes the form

$$\int_{\partial P} c^{\bullet}\xi\cdot\mathbf{n} = -\text{outflow}(\sigma(\mathbf{m}),P) + \int_{\partial E} V\mathbf{t}(\mathbf{m})\cdot\nu + o(\delta) \quad . \qquad (5.27)$$

Combining (5.26), (5.27), and (5.2),

$$\begin{aligned}
\{\Sigma\int_{P_i}\psi_i(c) + \int_E\sigma(\mathbf{m}) + o(\delta)\}^{\bullet} &= \text{outflow}(\sigma(\mathbf{m}),P) \leq \\
&- \Sigma\int_{(\partial P)_i}\mu_i\mathbf{h}_i\cdot\mathbf{n} + \int_{\partial E} V\mathbf{t}(\mathbf{m})\cdot\nu + o(\delta) \quad . \qquad (5.28)
\end{aligned}$$

which is the desired result, the formal equivalency of the two forms of the second law. (Note that each of the integrals in (5.28) is either $0(1)$ or $0(\delta)$.)

Acknowledgement.

This work was supported by the Army Research Office and by the National Science Foundation.

References.

[1878] Gibbs, J. W., On the equilibrium of heterogeneous substances, Trans. Connecticut Acad. **3**. 108-248 (1876). Reprinted in: The Scientific Papers of J. Willard Gibbs, Vol. 1, Dover, New York (1961).

[1893] van der Waals, J. D., The thermodynamic theory of capillarity under the hypothesis of a continuous variation of density (in Dutch), Verhandel. Konink, Akad. Weten. Amsterdam (Sec. 1) Vol. 1, No. 8.

[1901] Korteweg, D. J., Sur la forme que prennent les equations des mouvement des fluides si l'on tientcompte des forces capillaires par des variations de densite, Arch. Neerl. Sci. Exactes Nat. Ser. 11, **6**. 1-24.

[1951] Herring, C., Surface tension as a motivation for sintering, The Physics of Powder Metallurgy (ed. W. E. Kingston) McGraw-Hill, New York.

[1958] Cahn, J. W. and J. E. Hilliard, Free energy of a nonuniform system. **I**. Interfacial free energy, J. Chem. Phys. **28**, 258-267.

[1958] Frank, F. C., On the kinematic theory of crystal growth and dissolution processes, Growth and Perfection of Crystals (eds. R. H. Doremus, B. W. Roberts, D. Turnbull) John Wiley, New York.

[1961] Cahn, J. W., On spinodal decomposition, Act. Metall. **9**, 795-801.

[1962] Cahn, J. W., On spinodal decomposition in cubic crystals, Act. Metall. **10**, 179-183.

[1963] Coleman, B. D. and W. Noll, The thermodynamics of elastic materials with heat conduction and viscosity, Arch. Rational Mech. Anal. **13**, 167-178.

[1963] Mullins, W. W. and R. F. Sekerka, Morphological stability of a particle growing by diffusion or heat flow, J. Appl. Phys. **34**, 323-329.

[1964] Mullins, W. W. and R. F. Sekerka, Stability of a planar interface during solidification of a dilute binary alloy, J. Appl. Phys. **35**, 444-451.

[1964] Voronkov, V. V., Conditions for formation of mosaic structure on a crystallization front [in Russian], Fizika Tverdogo Tela **6**, 2984-2988. English Transl. Sov. Phys. Solid State **6**, 2378-2381 (1965).

[1971] Cahn, J. W. and J. E. Hilliard, Spinodal decomposition: a reprise, Act. Metall. **19**, 151-161.

[1972] Hoffman, D. W. and J. W. Cahn, A vector thermodynamics for anisotropic surfaces - 1. Fundamentals and applications to plane surface junctions, Surface Sci. **31**, 368-388.

[1974] Cahn, J. W. and D. W. Hoffman, A vector thermodynamics for anisotropic surfaces - 2. curved and faceted surfaces, Act. Metall. **22**, 1205-1214.

[1984] Carr, J., Gurtin, M. E., and M. Slemrod, Structured phase transitions on a finite interval, Arch. Rational Mech. Anal. **86**, 317-351.

[1986] Gurtin, M. E., On the two-phase Stefan problem with interfacial energy and entropy, Arch. Rational Mech. Anal. **96**, 199-241.

[1988] Fonseca, I., Interfacial energy and the Maxwell rule, Arch. Rational Mech. Anal., forthcoming.

[1988a] Gurtin, M. E., Toward a nonequilibrium thermodynamics of two phase materials, Arch. Rational Mech. Anal. **100**, 275-312.

[1988b] Gurtin, M. E., Multiphase thermomechanics with interfacial structure. 1. Heat conduction and the capillary balance law, Arch. Rational Mech. Anal., forthcoming.

[1988c] Gurtin, M. E., On a nonequilibrium thermodynamics of capillarity and phase, Q. Appl. Math., forthcoming.

[1988] Pego, R. L., Front migration in the nonlinear Cahn-Hilliard equation, forthcoming.

INSTABILITIES IN SHEAR FLOW OF VISCOELASTIC FLUIDS WITH FADING MEMORY *

Bradley J. Plohr

Computer Sciences Department
and
Center for the Mathematical Sciences
University of Wisconsin–Madison
Madison, WI 53706

1. Introduction

For certain models of viscoelastic fluids with fading memory, classical steady channel flow does not exist beyond a maximal wall shear stress. This occurs when the shear stress for steady flow decreases with strain. For generalized Newtonian models of viscoelastic flow, such a decrease implies that the flow is unstable. Because of this example, a constitutive relation that exhibits a maximal wall shear stress is regarded as defective.

We report on work showing that, contrary to this intuition, such models correctly describe the experimentally observed "spurt" phenomenon: exceeding this critical stress results in a large increase in volumetric flow rate. The transition to spurt flow is analogous to a dynamically generated phase transition.

To analyze this phenomenon, we derive a system of conservation laws that govern the flow; these equations take the form of gas dynamics with relaxation terms. We solve the Riemann problem for this non–strictly-hyperbolic system, and incorporate this solution into the random choice method. Numerical simulation of channel flow, with the maximal wall shear stress exceeded, shows that a discontinuity forms at the wall, allowing the fluid to "slip"; no steady state exists. However, when a small Newtonian viscosity is included in the model, a "slip layer" forms and the flow approaches a discontinuous steady state.

* Supported by the U. S. Department of Energy, the National Science Foundation, and the U. S. Army Research Office.

2. Viscoelastic Fluids with Fading Memory

The motion of a fluid is defined by the spatial position $\vec{x}(\vec{X},t)$ at time t of each fluid particle $\vec{X}$. The velocity of a particle is then $\vec{v} = \partial\vec{x}/\partial t$. The deformation of the fluid is characterized by the deformation gradient $F = \partial\vec{x}/\partial\vec{X}$. Alternatively, the deformation undergone by the fluid from time s to time t may be specified using the relative deformation gradient $F_s(t) = F(t)F(s)^{-1} = \partial\vec{x}(t)/\partial\vec{x}(s)$. We will work in the Eulerian picture, so all fluid variables are regarded as functions of the spatial position $\vec{x}$ at time t.

When the flow is incompressible and isothermal, the equations governing the evolution of F and $\vec{v}$ are the continuity equation,

$$\dot{F} = (\nabla\vec{v})F \ , \tag{2.1}$$

and the conservation of momentum,

$$\rho\dot{\vec{v}} = \nabla\cdot\sigma \ . \tag{2.2}$$

Here ρ is the mass density, σ is the Cauchy stress, and a dot denotes the convective time derivative; also, the components of $\nabla\vec{v}$ are $(\nabla\vec{v})^i{}_j = \partial v^i/\partial x^j$. We will write $\sigma = -pI+\Sigma$, where p is an isotropic pressure that maintains incompressibility; Σ is the extra stress.

The response characteristics of the fluid are embodied in the constitutive relation for the stress. For viscoelastic fluids with fading memory, these relations specify the extra stress as a functional of the deformation history of the fluid. Many sophisticated constitutive models have been devised; see Ref. [1] for a survey.

In the present work we focus on the Johnson-Segalman model [5] as a prototype for general constitutive models. This model accounts for non-affine deformation of Gaussian networks by introducing a parameter a, $-1 \leq a \leq 1$, that enters a nonlinear generalization of the classical Maxwell model:

$$\Sigma = \int_{-\infty}^{t} m(t-s)\left\{\bar{F}_s(t)\bar{F}_s(t)^T - I\right\} ds \ , \tag{2.3}$$

where $\bar{F}_s$ satisfies

$$\dot{\bar{F}}_s = -R_a^T \bar{F}_s \ , \qquad \bar{F}_s(s) = I \ . \tag{2.4}$$

Here $R_a = \omega - ad$, with $d = \frac{1}{2}\left[\nabla\vec{v} + (\nabla\vec{v})^T\right]$ and $\omega = \frac{1}{2}\left[\nabla\vec{v} - (\nabla\vec{v})^T\right]$ being the deformation rate and vorticity tensors. When $a = 1$, $\bar{F}_s = F_s$ and the Johnson-Segalman model reduces to the Maxwell model.

The function m is a memory kernel, defined on $[0, \infty)$, that we assume to be smooth, nonnegative, and integrable. A typical example is

$$m(s) = \mu\lambda e^{-\lambda s} \ , \tag{2.5}$$

where μ is a shear modulus and λ is a relaxation rate. With this choice of memory kernel, the Johnson-Segalman model may be formulated as the differential equation

$$\dot{\Sigma} + \Sigma R_a + R_a^T \Sigma = 2a\mu d - \lambda\Sigma \tag{2.6}$$

for the extra stress.

Constitutive relations such as Eq. 2.6 exhibit a mixture of elastic and viscous behavior. This may be seen heuristically as follows. In the short relaxation-time limit, $\lambda \to 0$, Eq. 2.6 shows that an objective time derivative of Σ is proportional to the deformation rate: $\dot{\Sigma} \sim 2a\mu d$. This is characteristic of elastic behavior, and leads to the interpretation of $a\mu$ as a shear modulus. By contrast, when $\lambda, \mu \to \infty$ with μ/λ fixed, $\Sigma \sim 2(a\mu/\lambda)d$; thus, the model displays viscous behavior with $a\mu/\lambda$ being the Newtonian shear viscosity coefficient.

As shown in Ref. [5], the constitutive relation for the stress derives from the energy

$$\mathcal{E} = \frac{1}{2a\rho}\int_{-\infty}^{t} m(t-s)\,\mathrm{Tr}\left\{\bar{F}_s(t)\bar{F}_s(t)^T - I\right\} ds \ . \tag{2.7}$$

The energy satisfies the conservation principle

$$\rho\left(\tfrac{1}{2}\vec{v}\cdot\vec{v} + \mathcal{E}\right)^{\cdot} = \nabla\cdot(\vec{v}\cdot\sigma) - \rho r \ , \tag{2.8}$$

where

$$r = -\frac{1}{2a\rho}\int_{-\infty}^{t} m'(t-s)\,\mathrm{Tr}\left\{\bar{F}_s(t)\bar{F}_s(t)^T - I\right\} ds \tag{2.9}$$

represents the rate of dissipation of energy.

3. Shear Flow

Interesting properties of viscoelastic constitutive relations are exhibited in simple planar shear flow. With the flow aligned along the y-axis, the deformation gradient F takes the form

$$F = \begin{pmatrix} 1 & 0 \\ \epsilon & 1 \end{pmatrix} , \tag{3.1}$$

where the strain ϵ depends only on the spatial coordinate x and the time t. In particular, the flow is incompressible; we assume ρ to be constant throughout the fluid. Let v denote the y-component of $\vec{v}$ and let σ denote the shear stress $\Sigma^{xy} = \Sigma^{yx}$. The variables v, σ, Σ^{xx}, and Σ^{yy} also depend only on x and t, while the pressure p takes the form $p(x,y,t) = p_0(x,t) - f(t)y$. The quantity f corresponds to a pressure gradient driving the flow.

In this notation, the continuity equation, Eq. 2.1, reduces to

$$\epsilon_t - v_x = 0 , \tag{3.2}$$

and the y-component of conservation of momentum, Eq. 2.2, becomes

$$\rho v_t - \sigma_x = f . \tag{3.3}$$

Furthermore, the x-component of the conservation of momentum equation determines p_0 through the equation $p_x = (\Sigma^{xx})_x$.

The stress components are expressed in terms of the history of the strain through the constitutive relation, Eq. 2.3. As is easily determined,

$$\sigma = \frac{a}{\alpha}\int_{-\infty}^{t} m(t-s)\sin\left(\alpha[\epsilon(t)-\epsilon(s)]\right) ds , \tag{3.4}$$

where $\alpha = \sqrt{1-a^2}$. Similarly, $\Sigma^{xx} = -\frac{1}{2}(1-a)\mathcal{N}_1$ and $\Sigma^{yy} = \frac{1}{2}(1+a)\mathcal{N}_1$, where $\mathcal{N}_1 = \Sigma^{yy} - \Sigma^{xx}$ is the principal normal stress difference:

$$\mathcal{N}_1 = \frac{2a}{\alpha^2} \int_{-\infty}^{t} m(t-s) \{1 - \cos(\alpha[\epsilon(t) - \epsilon(s)])\} \, ds \ . \tag{3.5}$$

Also, the energy is given by $\rho\mathcal{E} = \frac{1}{2}\mathcal{N}_1$. Notice that $\mathcal{N}_1 \geq 0$.

In the special case when m is given by Eq. 2.5, Eqs. 3.2–5 imply the differential equations

$$\rho v_t - \sigma_x = f \ , \tag{3.6}$$

$$\sigma_t - \left[a\mu - \alpha^2 \rho\mathcal{E}\right] v_x = -\lambda\sigma \ , \tag{3.7}$$

$$\rho\mathcal{E}_t - \sigma v_x = -\lambda\rho\mathcal{E} \ . \tag{3.8}$$

Several mathematical results are known for this quasilinear system; see Ref. [10] for discussion and references.

- The system is strictly hyperbolic if and only if

$$\rho\mathcal{E} < \frac{a\mu}{\alpha^2} \ , \tag{3.9}$$

and the change of type has been associated to certain instabilities.

- For smooth, hyperbolic initial data with sufficiently small variation, a classical solution exists for all time.
- Nevertheless, smooth initial data exist for which the solution develops singularities within a finite time. As the critical time is approached, the spatial derivatives of v and σ become unbounded.

Thus, the fading memory acts as a weak dissipative mechanism: the source terms in the equations serve to counteract the formation of singularities from sufficiently smooth data. When singularities do form, however, Eqs. 3.6–8 are no longer valid because the products of distributions $\mathcal{E}v_x$ and σv_x are ill-defined. Instead, the physical conservation equations, interpreted in the sense of distributions, govern the flow.

4. Equivalent Conservation Form

Eqs. 3.2–3 and 3.4 express physical conservation principles that remain valid past "blow-up." They are not standard conservation laws, however, because the stress is given by a convolution integral. To solve these equations, we first derive an equivalent system of equations that is more amenable to analysis.

This system is obtained by introducing $z \geq 0$ and θ through the equations

$$z \cos(\alpha\theta) = m * \cos(\alpha\epsilon) \ , \tag{4.1}$$

$$z \sin(\alpha\theta) = m * \sin(\alpha\epsilon) \ , \tag{4.2}$$

where we use the convolution notation

$$m * \phi(\epsilon) = \int_{-\infty}^{t} m(t-s)\phi(\epsilon(s))ds \ . \tag{4.3}$$

Notice that z and θ are continuous in t for almost every x, assuming that ϵ is of bounded variation. Therefore, in calculating the distributional time derivatives of z and θ, we may legitimately multiply by functions of z and θ. Simple algebra shows that

$$z_t = m(0)\cos(\alpha[\epsilon-\theta]) + \cos(\alpha\theta)\ m' * \cos(\alpha\epsilon) + \sin(\alpha\theta)\ m' * \sin(\alpha\epsilon) \ , \tag{4.4}$$

$$\alpha\, z\, \theta_t = m(0)\sin(\alpha[\epsilon-\theta]) - \sin(\alpha\theta)\ m' * \cos(\alpha\epsilon) + \cos(\alpha\theta)\ m' * \sin(\alpha\epsilon) \ . \tag{4.5}$$

Furthermore, the stress components are given as functions of ϵ, z, and θ:

$$\sigma = a\alpha^{-1}\, z \sin(\alpha[\epsilon-\theta]) \ , \tag{4.6}$$

$$\mathcal{N}_1 = 2a\alpha^{-2}\left\{\mu_{\text{tot}} - z\cos(\alpha[\epsilon-\theta])\right\} \ , \tag{4.7}$$

with $\mu_{\text{tot}} = \int_0^\infty m(s)ds$.

In this way, the constitutive relation 3.4 is transformed into Eq. 4.6 together with the conservation laws 4.4–5. The convolution integral that appeared in the definition of σ has been removed; in its place is a pair of evolution equations for the "internal variables" z and θ. Because the stress depends on ϵ and θ only through the combination $\tau = \epsilon - \theta$, θ may be regarded as the unrecoverable, or "viscoplastic," part of the strain, while τ is the "elastic" strain.

5. Nonlinear Waves

Let us now restrict to the case of a single relaxation mode: $m(s) = \mu\lambda \exp(-\lambda s)$. If we eliminate θ in favor of $\tau = \epsilon - \theta$ by combining Eq. 4.5 with Eq. 3.2, then Eq. 3.2 for ϵ decouples from the system. Therefore, we obtain the conservation laws

$$\tau_t - v_x = -\mu\lambda\alpha^{-1}\, z^{-1} \sin(\alpha\tau) \ , \tag{5.1}$$

$$\rho v_t - \sigma(\tau, z)_x = f \ , \tag{5.2}$$

$$z_t = -\lambda \{z - \mu \cos(\alpha\tau)\} \tag{5.3}$$

together with the constitutive relation

$$\sigma(\tau, z) = a\alpha^{-1}\, z \sin(\alpha\tau) \ . \tag{5.4}$$

In the present isothermal approximation, energy is not conserved in discontinuous flows. Rather, the energy equation, Eq. 2.8, serves as an "entropy" condition for weak solutions:

$$\rho \left[\tfrac{1}{2}v^2 + \mathcal{E}(\tau, z)\right]_t - [\sigma(\tau, z)v]_x \leq vf - \lambda\rho\mathcal{E}(\tau, z) \tag{5.5}$$

where

$$\rho\mathcal{E}(\tau, z) = a\alpha^{-2}\, \{\mu - z\cos(\alpha\tau)\} \ . \tag{5.6}$$

Except for the presence of source terms, Eqs. 5.1–3 are identical to those for Lagrangian, homentropic gas dynamics, with τ/ρ identified as specific volume, $-\sigma$ as pressure, and z as a function of the entropy. Corresponding to the fundamental thermodynamic identity is

$$\rho\, d\mathcal{E} = \sigma\, d\tau + \rho T\, dz \ , \tag{5.7}$$

where $T = -\rho^{-1} a\alpha^{-2} \cos(\alpha\tau)$ plays the role of "temperature."

However, the equation of state, Eq. 5.6 or Eq. 5.4, would be rather peculiar for a gas; *cf.* Ref. [8]. The physical state of the fluid is unchanged if τ is replaced by $\tau + 2\pi/\alpha$, so state space is cylindrical; for convenience, we require that $-\pi/2 \leq \alpha\tau < 3\pi/2$, along with

$z \geq 0$. The wave speeds are $\pm c$ and 0, where $\rho c^2 = \partial\sigma/\partial\tau = az\cos(\alpha\tau)$ defines the sound speed c. Therefore the system is strictly-hyperbolic if $-\pi/2 < \alpha\tau < \pi/2$, whereas there is a complex-conjugate pair of purely imaginary wave speeds when $\pi/2 < \alpha\tau < 3\pi/2$. The wave mode with zero speed is linearly degenerate, and the sound modes with speed $\pm c$ are genuinely nonlinear in the hyperbolic region except at $\tau = 0$.

Despite the peculiarities of the system, we have determined completely the structure of scale-invariant nonlinear waves. Such a wave consists of a sequence of elementary scale-invariant waves, either centered discontinuities or rarefaction waves, connecting constant states on the left and right. Discontinuities are required to satisfy Liu's generalization of Oleĭnik's entropy condition, which guarantees that the energy dissipation inequality 5.5 holds. This admissibility condition is equivalent to requiring shock waves to have viscous profiles: with Eq. 5.2 augmented by the Newtonian viscosity term $(\eta v_x)_x$, admissible shock waves arise as limits as $\eta \to 0$ of travelling wave solutions.

The wave structure is conveniently depicted with a wave curve, the locus of states $U = (\tau, v, z)$ on the right for a fixed state $U_L = (\tau_L, v_L, z_L)$ on the left. Only z may change across waves with zero speed, so these wave curves are trivial. On the other hand, z remains constant across waves corresponding to the characteristic families with speeds $\pm c$, so we suppress the z component of U when drawing wave curves. Fig. 1 shows the wave curve of the $-c$ family for a representative initial state U_L; the wave curve of the opposite family is obtained by reflection through the vertical line $v = v_L$. The figure was produced using a computer program that constructs the wave curves for general systems of conservation laws [4]. Of particular interest is the manner in which the wave curve corresponding to the $-c$ characteristic family continues past the boundary of the elliptic region with waves that contain rarefaction and shock waves of the $+c$ family; this is possible because the wave speeds coincide at the elliptic region.

With the wave curves known, Riemann initial-value problems may be solved. We have written a computer program that solves Riemann problems, and have incorporated it into

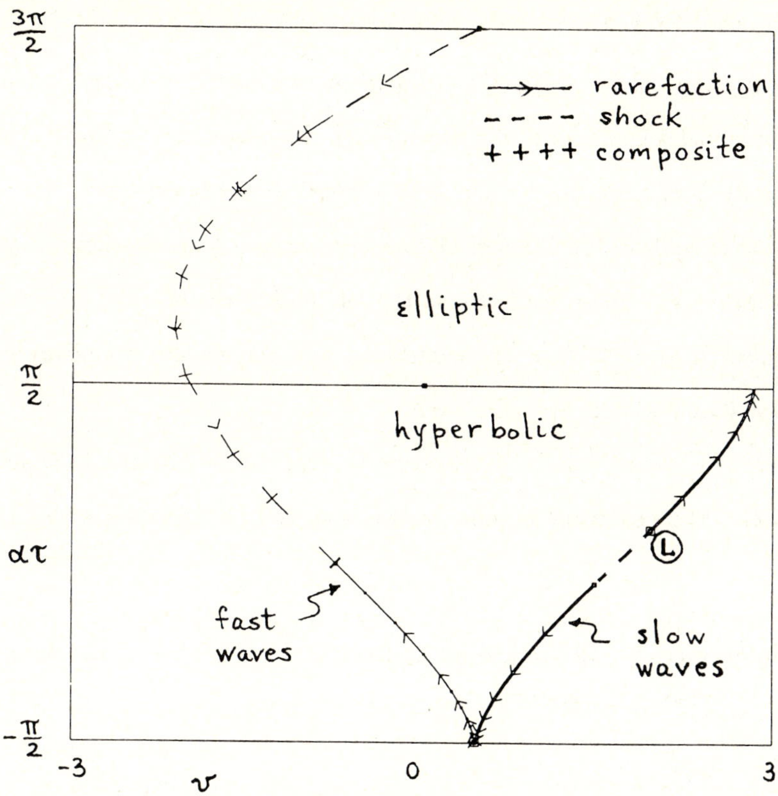

Fig. 1: The wave curve for a typical U_L. Solid curves correspond to rarefaction waves, dashed curves to shock waves, and dashed curves with crossbars to composite waves. Doubled curves represent the $-c$ family, while single curves represent the $+c$ family.

the Glimm-Chorin random choice method. We refer to Ref. [9] for a detailed discussion of this method. The Riemann problem solver can also be used in enhanced computational methods for two-dimensional flow problems, such as the front tracking method [2].

6. Steady Channel Flow

As an application of the results just described, we have investigated a flow of a Johnson-Segalman viscoelastic fluid in a channel [7]. In particular, we have studied the approach to steady flow, since this corresponds to rheological experiments.

Let the vertical walls of the channel be located at $x = \pm\frac{1}{2}w$. The flow is symmetric about the centerline ($x = 0$), so we focus on the left half of the channel, $-\frac{1}{2}w \le x \le 0$. Symmetry requires that $v_x = 0$ at the centerline, and the no-slip boundary condition imposes that $v = 0$ at the wall.

Steady shear flow occurs when the velocity v is independent of time. In this case, the strain rate $\kappa = \epsilon_t = v_x$ is constant in time, so that the stress σ coincides with

$$\sigma_{\mathbf{steady}}(\kappa) = \frac{a}{\alpha}\int_0^\infty m(s)\sin(\alpha\kappa s)\,ds\ . \tag{6.1}$$

The steady solution may be determined by integrating, in turn, the ordinary differential equations $-\sigma_{\mathbf{steady}}(\kappa)_x = f$ and $v_x = \kappa$, with the boundary conditions that $\kappa = 0$ at $x = 0$ and $v = 0$ at $x = -\frac{1}{2}w$.

Steady solutions may fail to exist if $\sigma_{\mathbf{steady}}(\kappa)$ does not vary monotonically with κ. For example, when $m(s) = \mu\lambda\exp(-\lambda s)$,

$$\sigma_{\mathbf{steady}}(\kappa) = \frac{a\mu}{\alpha}\,\frac{\alpha\kappa/\lambda}{1+(\alpha\kappa/\lambda)^2} \tag{6.2}$$

attains a maximum $\sigma_{\mathbf{max}} = \frac{1}{2}a\mu/\alpha$ when $\alpha\kappa/\lambda = 1$. Classical steady flow in a channel of width w cannot exist if the pressure gradient f exceeds a critical value f_{crit} given by $\sigma_{\mathbf{max}} = \frac{1}{2}wf_{\mathrm{crit}}$. Notice that this phenomenon is not associated with loss of hyperbolicity, since $\rho\mathcal{E} = \frac{1}{2}a\mu/\alpha^2$ at the stress maximum, well within the bound 3.9.

Traditionally, the non-existence of classical steady solutions caused by a shear stress maximum is regarded as a defect of the constitutive law. This conclusion is based on intuition appropriate for generalized Newtonian models of viscoelastic fluids. Shear flow for such a fluid is governed by the single equation

$$\rho v_t - [\eta(v_x)v_x]_x = f\ , \tag{6.3}$$

corresponding to having a viscosity coefficient η that depends on strain-rate. In a flow regime where $\eta(\kappa)\kappa$ decreases with strain-rate κ, however, Eq. 6.3 has the character of a backward heat equation, which suffers from the Hadamard instability. Therefore, for generalized Newtonian fluids, $\eta(\kappa)\kappa$ must increase with κ in a physically stable steady solution.

The Johnson-Segalman model has the same steady solutions as does a generalized Newtonian fluid with $\eta(\kappa)\kappa = \sigma_{\mathrm{steady}}(\kappa)$, so one might think that it exhibits the same instability in regions where σ_{steady} decreases. This conclusion is not warranted, however, because the system of equations maintains its evolutionary character even near the shear stress maximum.

A question remains: what is the solution if the critical pressure gradient f_{crit} is exceeded? To answer this we must solve the time-dependent equations, allowing for the possible formation of singularities.

7. Slip and Spurt Phenomena

We have simulated numerically a simple experiment for a fluid governed by the system 5.1–4. Without loss of generality, units are chosen so that $a\mu = 1$, $\lambda = 1$, and $\rho = 1$, and the fluid variables are scaled so that a disappears from the equations. Given this choice, the channel thickness w and the pressure gradient f (which depends on time) are the essential parameters defining the flow. In the numerical experiment, the channel width is $w = 1$, so that the wall of the channel is located at $x = -\frac{1}{2}$ while the center is at $x = 0$. The flow is initially in the classical steady state corresponding to the critical pressure gradient f_{crit}; then the pressure gradient is increased abruptly to the super-critical value $1.2 f_{\mathrm{crit}}$.

The result of this calculation using the random choice method is shown in Fig. 2. The fluid velocity v is plotted *vs.* position x at successive time intervals; generally the velocity increases with time. During the early stages of the experiment, the flow settles into a quasi-steady state. This effect is especially evident in a plot of the centerline velocity as a function of time, and it is more pronounced when the channel width w is smaller [7]. Eventually, however, a thin layer develops at the wall in which the velocity rises to a value that is nearly constant across the channel. For practical purposes, the fluid has broken free from the wall and is accelerating uniformly under the applied pressure gradient; thus the fluid "slips."

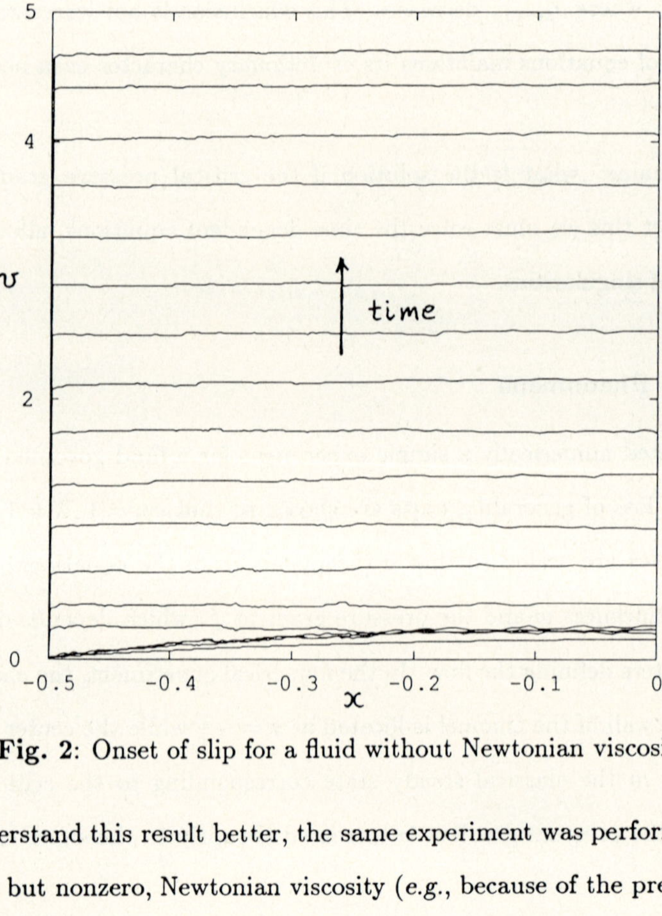

Fig. 2: Onset of slip for a fluid without Newtonian viscosity.

To understand this result better, the same experiment was performed for a fluid that has a small, but nonzero, Newtonian viscosity (*e.g.*, because of the presence of a solvent). For such a fluid, the term $(\eta v_x)_x$ is added to the right-hand-side of Eq. 5.2. So long as $0 \leq \eta < 1/9$ (in dimensionless units), the steady shear stress, plotted *vs.* strain, still exhibits a maximum.

Fig. 3 shows the results for $\eta = 0.01$, as calculated using the Lax-Wendroff method with Tyler artificial viscosity. Now the slip layer is much thicker, with its outer boundary corresponding to a discontinuity in the strain rate v_x. The solution approaches a steady state in which v_x is discontinuous but the total stress $\sigma_{\mathbf{steady}}(v_x) + \eta v_x$ is continuous.

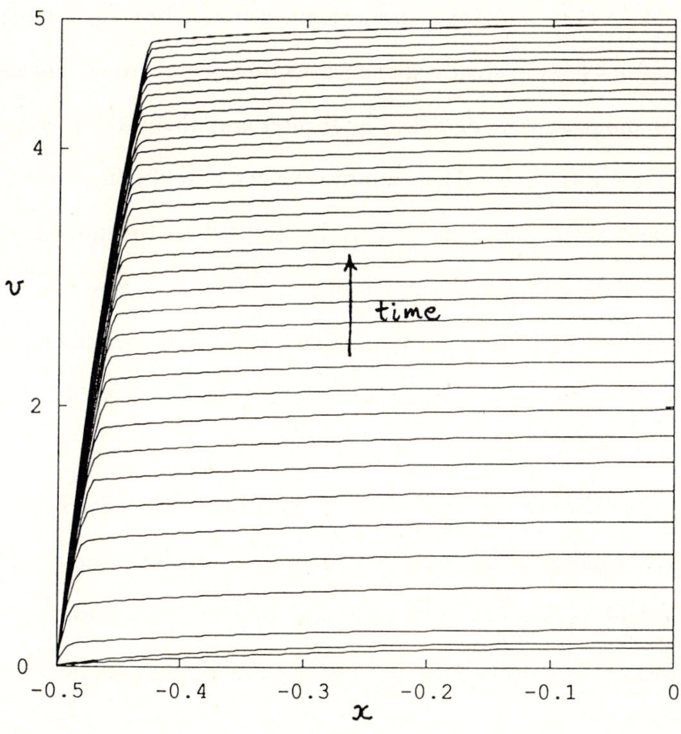

Fig. 3: Onset of spurt for a fluid with Newtonian viscosity.

Further study of channel flow for a Johnson-Segalman fluid yields several observations [7].

• The nominal shear rate (based on volumetric flow rate), plotted against the nominal shear stress (based on the pressure gradient), shows an abrupt increase at a critical point, *i.e.*, "spurt."

• The thickness of the slip layer (or, equivalently, the stress at which the strain-rate discontinuity occurs) exhibits hysteresis as a function of the loading.

• There are three essential time scales to the flow problem, corresponding to processes governed by Newtonian viscosity, by wave propagation, and by relaxation. In some situations this leads to a quasi-steady supercritical flow that persists for a time orders of magnitude longer than the wave propagation scale.

There are two essential features responsible for spurt and hysteresis phenomena in viscoelastic models: (1) a non-monotone stress-strain relation in steady flow; and (2) a dynamical mechanism for selecting steady shock waves that are physical. For example, the spurt and hysteresis phenomena were analyzed by Hunter and Slemrod [3] for a related, but simpler, model of a viscoelastic fluid. Their analysis shows how the first two observations may be understood, and we expect that it may be applied to the present model. We emphasize, however, that the Johnson-Segalman model differs in an essential way from the model considered by Hunter and Slemrod: it remains hyperbolic even in the range where the stress decreases with strain, as is desirable for equations that describe evolution of a physical system.

Spurt and hysteresis offers an explanation of certain observations in rheological experiment [11]. Quantitative agreement between experiment [11] and Johnson-Segalman model has been verified in Ref. [6]. Therefore, we have confidence that the interesting mathematical structure of models of viscoelastic flow reflects observable physical phenomena.

Acknowledgements

The author thanks Profs. D. Malkus and J. Nohel for introducing him to viscoelastic flows, and for many discussions that contributed to the present work. He also thanks Profs. M. Johnson and T-P. Liu for their encouragement and useful comments.

References

1. R. Bird, R. Armstrong and O. Hassager, *Dynamics of Polymeric Liquids*, John Wiley and Sons, New York, 1987.

2. J. Glimm, C. Klingenberg, O. McBryan, B. Plohr, D. Sharp and S. Yaniv, "Front Tracking and Two Dimensional Riemann Problems," *Adv. Appl. Math.* **6** (1985), pp. 259–290.

3. J. Hunter and M. Slemrod, *Phys. Fluids* **26** (1983), pp. 2345–2351.

4. E. Isaacson, D. Marchesin and B. Plohr, "Construction of Nonlinear Waves for Conservation Laws," in preparation, 1988.

5. M. Johnson and D. Segalman, "A Model for Viscoelastic Fluid Behavior which Allows Non-Affine Deformation," *J. Non-Newtonian Fluid Mech.* **2** (1977), pp. 255–270.

6. R. Kolkka, D. Malkus, M. Hansen, G. Ierley and R. Worthing, "Spurt Phenomena of the Johnson-Segalman Fluid and Related Models," Preprint, Michigan Technical University, Houghton, MI, 1987.

7. D. Malkus, J. Nohel and B. Plohr, "Spurt and Slip Phenomena in Shear Flows of Non-Newtonian Fluids," in preparation, 1988.

8. R. Menikoff and B. Plohr, "The Riemann Problem for Fluid Flow of Real Materials," Preprint LA-UR-88-49, Los Alamos National Laboratory, Los Alamos, NM, 1988.

9. B. Plohr, "Shockless Acceleration of Thin Plates Modeled by a Tracked Random Choice Method," *AIAA J.*, 1988. To appear.

10. M. Renardy, W. Hrusa and J. Nohel, *Mathematical Problems in Viscoelasticity*, Pitman Monographs and Surveys in Pure and Applied Mathematics, Vol. 35, Longman Scientific & Technical, Essex, England, 1987.

11. G. Vinogradov, A. Malkin, Yu. Yanovskii, E. Borisenkova, B. Yarlykov and G. Berezhnaya, "Viscoelastic Properties and Flow of Narrow Distribution Polybutadienes and Polyisoprenes," *J. Polymer Sci., Part A-2* **10** (1972), pp. 1061–1084.

SINGULARITIES OF THE ORDER PARAMETER IN CONDENSED MATTER PHYSICS

Maurice Kléman
Laboratoire de Physique des Solides (associé au CNRS),
Université de Paris-Sud, Bât. 510, F-91405 Orsay Cedex

Order parameter ; its amplitude. Landau theory of phase transitions

The order parameter, in the form it has been worked out by Landau [1], is first of all a thermodynamical variable, conjugated to a field, which describes at a macroscopic scale specific properties of condensed matter deriving from microscopic interactions. The standard example, which is developed in any text book of statistical physics, is the magnetization $\vec{M}(T)$ of a ferromagnetic medium, which describes, at a temperature T, the state of cooperative behaviour of the magnetic spins of the atoms. $\vec{M}(T)$ is the thermodynamic average $\Sigma <\vec{S}_i>$ of the spins per unit volume. The free energy density f of a ferromagnet can be written as :

$$f = f_0(T) + \frac{1}{2} a (T - T_c) M^2 + \frac{1}{3} B M^3 + \frac{1}{4} C M^4 + ... \qquad (1)$$

this expansion being strictly valid near the Curie temperature T_c, below which M is small, and above which the medium loses its ferromagnetic properties (for $T \geq T_c$, the minima of f are reached for $M \equiv 0$). f must be of course invariant under the action of any element of symmetry of the medium, but such a condition is trivially achieved when the order parameter is a constant vector, as it is here. More interesting is the remark, made by Landau himself, that a phase transition is second order only if the terms of order 3 vanish idendically in the free energy expansion (more precise group theoretical discussions of this theorem can be found in ref. 1 and 2). Therefore $B \equiv 0$ in eq. 1, since the ferromagnetic $\leftrightarrow$ paramagnetic transition is second order (fig. 1). This theorem originates in the fact that, at a second order phase transition, the group of

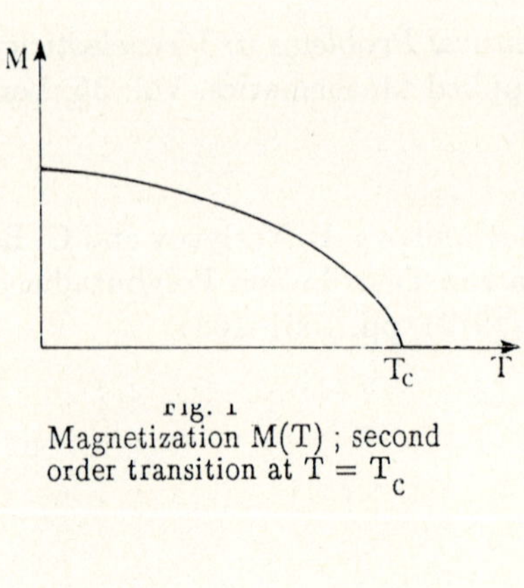

Fig. 1
Magnetization M(T) ; second order transition at $T = T_c$

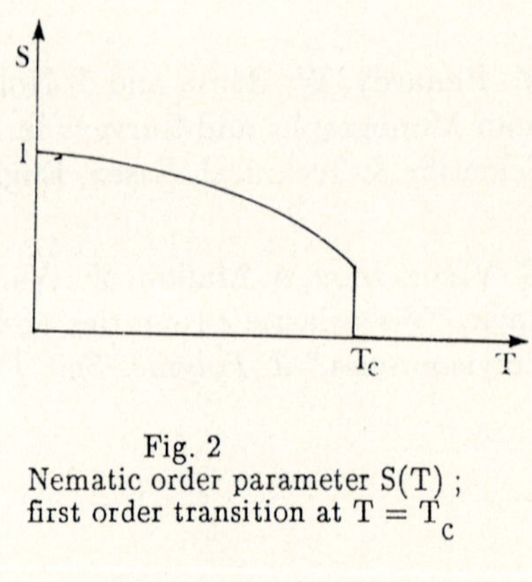

Fig. 2
Nematic order parameter S(T) ; first order transition at $T = T_c$

interesting properties of a variable order parameter are of a topological nature, and can be conceptualized under the categories of generalized phase and order parameter space. The non-linearities associated to the order parameter are, accordingly, of a topological nature.

<u>Order parameter ; its phase. Order parameter space ; singularities of the order parameter</u>

Let us start with the simplest example of a topologically non-trivial order parameter, the superfluid state of A-^{3}He. It is a condensate of boson particles, whose wave function

$$\psi = \psi_0(T) \exp i\zeta \qquad (4)$$

is constant in space in the ground state of the superfluid. This means that, not only $\psi_0(T)$, the amplitude of the order parameter, but also φ, the phase, is constant in space. If the medium is distorted at constant temperature,in first approximation, it is only the phase $\varphi(r)$ which varies. Any distortion has here the physical meaning of a superfluid velocity

$$\vec{v} \propto < \psi | \nabla | \psi > \propto \text{grad } \varphi \qquad (5)$$

Note that this velocity field is irrotational, and might carry vortices of the Kelvin type. The phase φ varies by an integral number n of times 2π, when completing a 2π turn around such vortex lines (fig. 3), since the order parameter ψ should not change value after such a circumnavigation. Therefore φ is multivalued around and singular on the line itself.

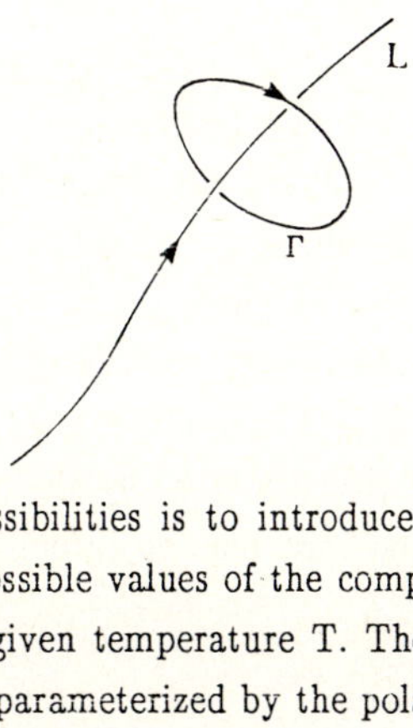

Fig. 3
Vortex line L and test circuit Γ ; the phase varies by $2\pi n$ around Γ

topological possibilities is to introduce a space whose points are in one-to-one correspondance with all the possible values of the complete order parameter [5]. Here this space V(A-^{3}He) is a <u>circle S^1</u> at a given temperature T. The radius of the circle can be taken proportional to $\psi_0(T)$, and its points parameterized by the polar angle φ, which varies by an amount equal to 2π around the circle. Then any circuit like Γ maps on n complete turns around this circle. Since we are interested in distortions at constant temperature, we do not want to consider further the radius of the circle, but only its topology. Clearly the <u>topological stability</u> of the vortex lines in ^{3}He, in particular the fact that the same invariant $2\pi n$ is carried by any circuit surrounding the line, stems in the fact that closed loops on S^1 cannot shrink continuously to zero. For if they could, it would mean that the invariant would vanish. All possible invariants are classified by :

$$\pi_1(S^1) = \mathbb{Z} \tag{6}$$

the first homotopy group of the circle. It will appear on other examples that it is always the topological properties of the space, not its metrical ones, which are of importance to describe the generalized phase of the order parameter. Non-linear properties are of course inherent in this approach of the order parameter singularities problem.

A less simple example, which will bring us immediately in the surprising novelties of the topological approach, arises with the case of the nematic phase. Here the complete order parameter includes an amplitude S(T) (see above), and a direction $\vec{n}$ of the ground state, $\vec{n}$ being an unit vector parallel to the optical axis, which is the mean (thermodynamic) orientation of some axis of anisotropy of the molecules. But $\vec{n}$ is identical to $-\vec{n}$, in the sense that the nematic organization is not polar, and has locally the symmetry of a cylinder. In other words, a convenient order parameter space is the projective plane P^2 (fig. 4) which is nothing else than a sphere with antipodal points identified. A sphere S^2 would be a suitable order parameter space for a ferromagnetic liquid or a ferromagnetic glass, in which the order parameter $\vec{M} \neq -\vec{M}$, so that two antipodal points $\vec{M}$ and $-\vec{M}$ would represent different local states.

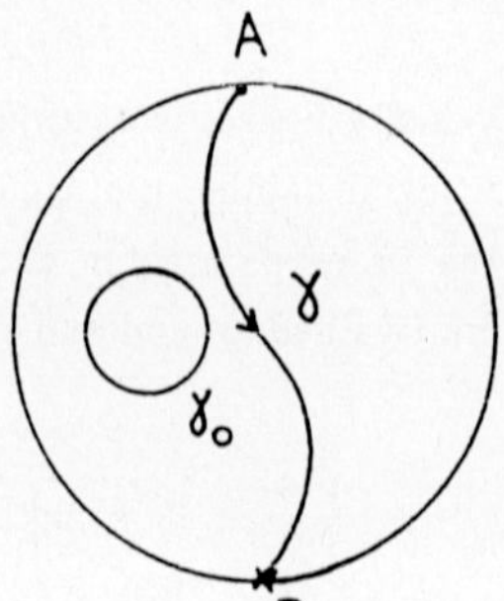

Fig. 4

The projective plane as an order parameter space for the nematic phase. See text.

Consider now a singular line L in some distorted nematic, and a circuit Γ surrounding this line. Map the values $\vec{n}$ of the director along Γ on P^2. Since Γ is drawn in the 'good' crystal, i.e. in a region where $\vec{n}$ is continuous, the image of Γ on P^2 is a closed line γ; by the same argument than above, we are interested in the homotopy class of γ. The first homotopy group of P^2

$$\pi_1(P^2) = \mathbb{Z}_2 \tag{7}$$

is an abelian group of order 2 with elements 1, a,

such that $a^2 = 1, \quad a = a^{-1}$.

Any closed loop of the type AA (fig. 4) is the image of a circuit Γ surrounding a singular line L. Locally, in a plane perpendicular to L, the configuration of the director $\vec{n}$ looks either as in fig. 5a, or in fig. 5b. It is possible to transform the configuration of fig. 5a into that one of fig. 5b, by a continuous rotation of the local directors, since both lines belong to the same class of homotopy. A nice representation of this continuous rotation has been done by Bouligand [6].

Any closed loop of the type γ_0 (fig. 4) is the image of a circuit Γ surrounding a line L which is not topologically stable (t.s.), i.e. which can exist in one of the two states drawn fig. 6, either singular, or non-singular. It is visible that it is possible to go continuously from the singular to the non-singular state by an 'escape in the 3rd dimension' of the director. Finally any two lines of the types drawn in fig. 5 can collapse and give a non-singular line, and then disappear, because of the relationship $a^2 = 1$.

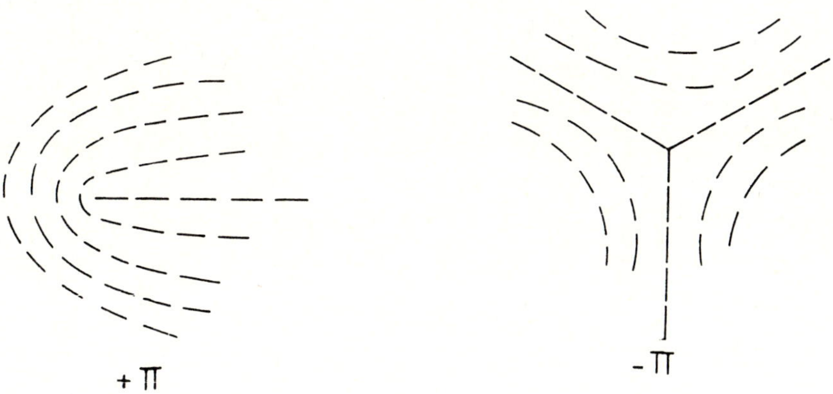

Fig. 5
Two possible configurations for a nematic singularity represented by the closed loop γ in fig. 4. In terms of the Volterra process (see below), the rotations Ω of the cut surface are opposite and equal to π.

Figure 6 tells us much more ; the escape in the 3rd dimension is possible either "up", or "down" ; if both possibilities happen on different segments along the line, these segments join at singular points, which can be classified equally well by homotopy considerations. The generalization of a closed loop surrounding a line, is, in the case of a singular point, a closed surface Σ homotopic to a sphere, surrounding the point. The image of Σ on P^2 in the 1-1 mapping is a closed 2-dimensional manifold σ, whose homotopy class belongs to the second homotopy group $\pi_2(P^2)$

$$\pi_2(P^2) = \mathbb{Z} \qquad (8)$$

The singular points which appear possibly on the line belong to two opposite classes, 1, and -1, say, of $\mathbb{Z}$ (in additive notation).

Starting from those results,it is possible to extend the topological theory of defects in three directions at least [7] :

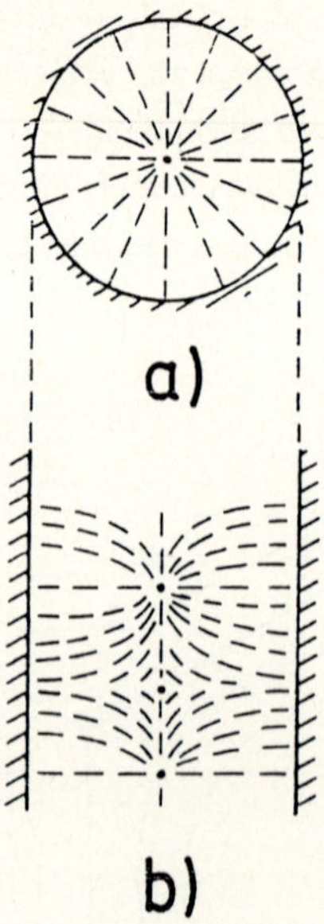

Fig. 6
A nematic singularity line in cylindrical geometry represented by γ_0 in fig. 4. The director can escape in the 3rd direction in either way, up or down. Singular points show up at the point of encounter of the two modes of escape.

- to give a physical content to $\pi_0(V)$ and $\pi_3(V)$, for any V . $\pi_0(V)$ classifies t.s. evalls ; $\pi_3(V)$ classifies t.s. "configurations", i.e. non singular, non-trivial, geometries of the order parameter which cannot relax continuously to the ground state.
- to understand the physical meaning of the relationships between successive homotopy groups which are provided by the exact homotopy sequence ; this remark has led to some significant results which concern the theory of defects in interaction with specimen boundaries.
- to define V for any ordered medium ; this problem has been solved in its full generality. $V = G/H$, the quotient of the euclidean group G : $R^3 \square 0(3)$ by the group of symmetry H of the ordered medium. In the case of a solid crystal, $H = T \square P$, the semi-direct product of the translation subgroup T (generally isomorphic to $\mathbb{Z} \times \mathbb{Z} \times \mathbb{Z}$) by the point group P. V is also known for all the mesomorphic crystals, and for media where the symmetry is of the gauge type (like in the above example of A-^{3}He).

At this point, it is worth mentionning the analogies and differences between the two known approaches to the theory of singularities, the other one being the so-called Volterra process, which is much older than the topological theory [8]. Volterra had in mind the study of the singularities of the (linear) equations of elasticity, when he devised his method of cutting a bounded surface in a unstrained medium, displacing the lips of the cut by some quantity $\vec{d}(\vec{M})$, removing (or adding) unstrained matter, and pasting finally this new matter along the lips of the cut, before letting the medium, transformed in this way, relax elastically (fig. 7). When the

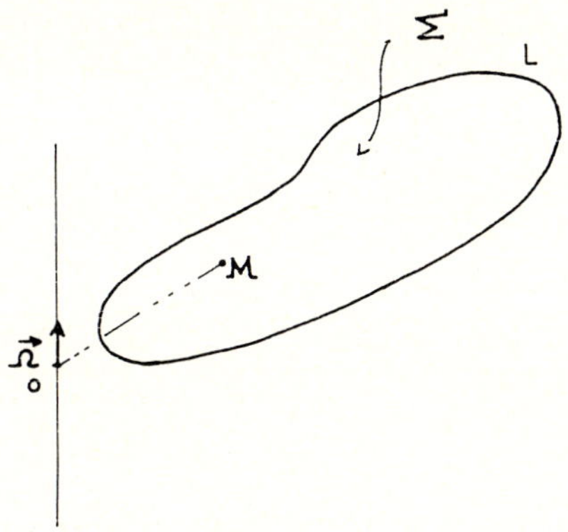

Fig. 7
The Volterra process

displacement $\vec{d}(\vec{M})$ is a rigid displacement, i.e. the sum of a translation $\vec{b}$ and a rotation Ω, the strain and stress fields thus created are singular along the line L bounding the cut surface, but are continuous through it (Weingarten theorem). The cylindrical region around the line (the so-called core of the line defect, of a few atomic distances wide), is the seat of stresses larger than the yield stress, i.e. it is a non-linear strained region. This non-linear behaviour of the strains and stresses, localized in the core, is the fundamental physical property of the types of singularities investigated by the Volterra process. The other character of a singularity evidenced by the Volterra process is its topological behaviour, i.e. the existence of an invariant $(\vec{b}, \Omega)$ which characterizes the line. In fact, it is the same invariant which is displayed by the topological theory when it is applied to the classification of line defects in crystalline solids, and, anyway, the invariants of the two theories are related. For example $\vec{b}$ is an element of the group of translation T, which is isomorphic to the first homotopy group of the 3-dimensional torms T^3, the order parameter space for the displacement field $\vec{u}(\vec{r})$; $\vec{u}(\vec{r})$ is indeed a valid 'generalized phase' for the order parameter of a solid, which consists as indicated above of the Fourier components of the mass density, whose complete form when the medium is distorted is $\rho_{\vec{k}} \exp i\, \vec{k}.\vec{u}$, in a first order approximation (since $\vec{r} \to \vec{r} + \vec{u}$). (fig. 8)

On the other hand, the topological approach does not show up directly the singularities of the stresses on the core, at least how they are related to the material constants. Its physical content is in this sense much more restricted, while it has a much wider range of application : to all kinds of order, and to all defects dimensionalities.

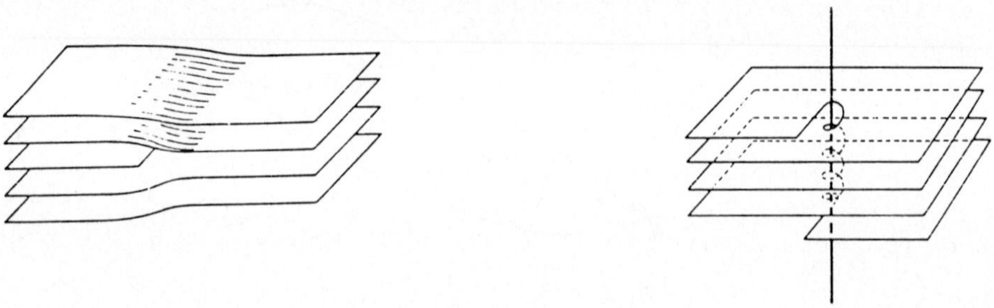

Fig. 8

Singularity lines of Volterra invariant ($\vec{b}$,O). We have drawn two possible realizations of such a singularity, the edge dislocation ($\vec{b}$ perpendicular to the line), and the screw dislocation ($\vec{b}$ parallel to the line)

Inhomogeneous ordered states ; geometrical frustration

With the discovery of blue phases and the evidence for local icosahedral order in simple liquids and amorphous metals, some impetus has been given in the last ten years to the crystallography of spaces of constant, non-zero, curvature, i.e. spheres and Lobachevski spaces. They are at the origin of new considerations on the singularities of the order parameter in media where the local order is non-euclidean, a phenomenon to which we reserve the name of geometrical frustration, since such a local order cannot extend homogeneously in space for only geometrical reasons.

The simplest example is the case of simple liquids and metallic glasses, for which the concept was first developped [9]. In these systems made of equal atoms, a condensed local state of small energy is icosahedral : one atom is surrounded by Z = 12 atoms in contact with the central one, but not in contact between them if their centers are exactly at the vertices of icosahedron. However, this geometry is locally denser than the more classical fcc or hcp packings ; this explains the stability of such a cluster, which is furthermore favoured by the vibrational entropy of the atoms on their sphere of coordination. But this symmetry cannot extend in space very far without introducing strong elastic distortions. Therefore either the medium crystallizes under the shape of an aperiodic crystal (quasi-crystal), or it is disordered. Is it possible to describe the structure of this disorder in more details ?

Icosahedral symmetry is frustrated in our usual euclidean, R^3 space ; but this frustration is relaxed in a curved space of constant positive curvature, the 3-d sphere S^3, where it is possible to build a 3-d 'crystal' of finite extent since S^3 is a finite space, made of 600 regular tetrahedra, 5 meeting around each edge, and 20 around each vertex, where they form an icosahedron. In Schläfli notations, this is the (3,3,5) polytope. The relationship between (3,3,5) and the disordered structure can be understood pictorially as follows : introduce disclination lines whose invariants Ω are multiple of $2\pi/5$. Such invariants respect the (3,3,5) symmetry. This process creates lines of vertices with coordination number $Z = 14, 15, 16...$ (fig. 9). Along these lines of defects, since

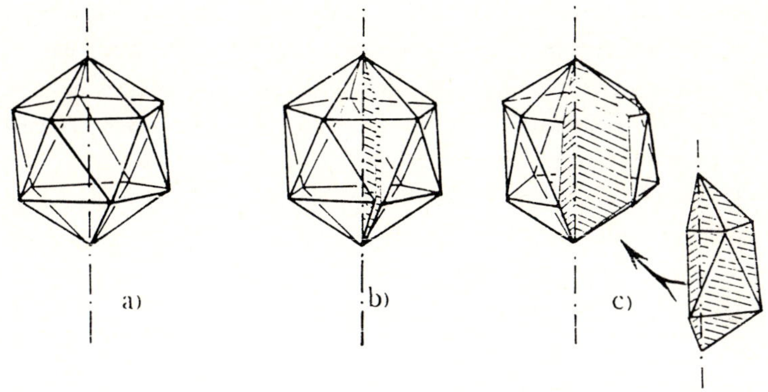

Fig. 9
a) icosahedral local order in a simple liquid or amorphous solid . Each vertex is the center of an atom. $Z = 12$
b) disclination on the icosahedral order ; the cut surface of the corresponding Volterra process
c) opening of the cut surface and introduction of extra-matter. $Z = 14$

extra matter has been introduced, the sphere is decurved. The purpose is of course to flatten S^3 completely. Most of the flattening can be achieved by lines of defects, with prescribed Ω's (as above), distributed at random. Of course, it will be necessary to introduce also deformations, since in fact the medium carry a slight negative curvature (instead of a large posivite one) around the lines where $Z = 14$, if the distances between atoms are not modified. The reader is invited to transpose this discussion to a 2-d sphere S^2 tiled with regular pentagons, and try to flatten it by a similar process.

Another way of understanding the relationship between the curved crystal and its flattened image is to consider a mapping from the curved space to the euclidean space which conserves lengths and angles. According to Cartan [10], such a mapping consists geometrically in a rolling along a line of the curved space on the flat one, without slipping. But such a mapping is successful only along the prescribed line, not in its vicinity. Furthermore a closed line scribed on the curved space rolls along an open path ; closure failure is directly related to the curvature, and hence to the density of disclinations which one has to introduce [11].

The final result is anyway a medium with unfrustrated domains (of some atomic distances wide) separated by defect lines, along which the coordination number Z is larger than 12 on the average. This model predicts quite reasonably the density of packing of an amorphous metal.

It is also remarkable that some tetrahedrally coordinated metallic alloys, of the so-called Frank and Kasper type [12], which are periodic, can be described in terms of disclination lines of the type above ($Z > 12$), periodically located in a matrix $Z = 12$ [13]. Some of those Frank and Kasper phases are 'approximants' of quasi-crystals [4]. It is therefore quite probable that quasi-crystals could be described within the same formalism, as mappings of curved crystals. In fact, a recent attempt has been made in this direction [14]. The curved crystal is in this case infinite and of negative curvature and is a tiling of triacontrahedra.

For the sake of completion, let us indicate that the case of blue phases, which are cubic periodic arrays of disclination lines in a chiral liquid crystal, can be treated within a quite similar formalism. As shown by Sethna [15], the corresponding unfrustrated crystal is the sphere S^3, with the lines of force of the molecules along a set of great circles. Some developments applying this new type of curved crystal to biological polymers (like DNA) and more generally locally oriented polymers, have been made in the recent years [16].

References

[1] Landau L.D. and Lifshitz E.M., Statistical Physics, Pergamon Press, 1958

[2] Boccara N., Symétries brisées, Hermann, Paris, 1976

[3] Kadanoff L.D., Physics, 1966, 2, 263
Toulouse G et Pfeuty P., Introduction au Groupe de renormalisation, Presses Universitaires de Grenoble, 1975

[4] for a recent restatement of the question, see 'Quasi-Crystalline Materials', Proceed. of the ILL/CODEST workshop, Ch. Janot and J.M. Dubois eds., World Scientific, Singapore, 1988

[5] Toulouse G. and Kléman M., J. de Physique Lett., 1976, 37, L-149

[6] Bouligand Y. in 'Physics of Defects', R. Balian, M. Kléman and J.P. Poirier eds., Les Houches School of Theoretical Physics, North-Holland, 1981, p. 665

[7] for a review on the topological theory of defects, see Michel L., Rev. Mod. Phys., 1980, 52, 617

[8] Friedel J., Dislocations Pergamon Press, London, 1964

[9] Kléman M. and Sadoc J-F., J. de Physique Lett., 1979, 40, 79

[10] Cartan E., Leçons sur la Géométrie des Espaces de Riemann, Gauthier-Villars, Paris, 1928

[11] Kléman M., J. de Physique Lett., 1983, 44, L-295

[12] Frank F.C. and Kasper J., Acta Cryst., 1958, 11, 184 ; 1959, 12, 489

[13] Sadoc J.F., J. de Physique Lett., 1983, 44, L-707
Nelson D., Phys. Rev., 1983, B 28, 5515

[14] Kléman M. and Ripamonti C., Comptes Rendus Acad. Sci. 1988, 307, 335

[15] Sethna J.P., Phys. Rev. Lett., 1983, 51, 2198

[16] Kléman M., J. de Physique Lett., 1985, 46, L-723 ; Physica Scripta, 1987, T 19, 565.

SWELLING AND SHRINKING OF POLYELECTROLYTE GELS

Ingo Müller
FB 9 - Physikalische Ingenieurwissenschaft
Technische Universität Berlin
Straße des 17. Juni 135, 1000 Berlin 12

Abstract

Polyelectrolyte gels in solution swell and shrink dramatically upon a change of temperature or of the acidity of the solvent. This paper shows that these phenomena can be understood as a shift in the delicate balance between three competing tendencies:

i.) osmotic pressure of the counter ions,
ii.) network elasticity,
iii.) van der Waals attraction.

1. Phenomena

Swelling and shrinking phenomena in gels are particularly pronounced in electrolytic gels, e.g. those consisting of polyacrylamide in water solution. Such gels were extensively investigated by Tanaka and his group, see [1], [2], [3]. The phenomenon of swelling or shrinking is provoked by either a change of temperature or by a change of the acidity of the solution as shown by Figure 1. The abscissae of the diagrams in that figure represent the volume fraction of the polymeric network in the gel so that the change of volume along the dashed lines is by a factor of 200 to 300. The purpose of this paper is to explain this phenomenon and to describe it. In this I follow the arguments presented by Morro & Müller [4].

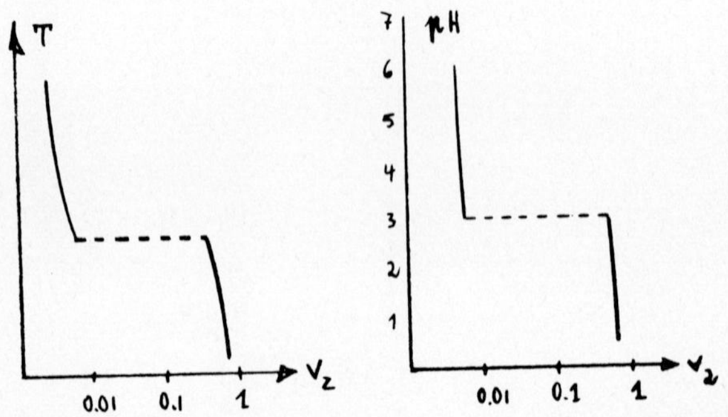

Figure 1: Collapse and swelling of a polyelectrolyte gel

2. Qualitative Explanation

The system under consideration consists of the gel and the solution bath in which the gel is immersed, see Figure 2. Some links in the chain molecules of the polymeric network of the gel are negatively ionized. The gel also contains the positive counterions and molecules of the solvent which surrounds the gel. There are three conflicting tendencies

i.) Osmotic Pressure, i.e. the tendency of the mobile counterions to spread throughout the system of gel and solvent. On the other hand strong electric forces prevent the counterions from leaving the gel. Therefore they can only spread by expanding the network.

ii.) Network elasticity, i.e. the tendency of the network to assume a small volume so that the long chain molecules can increase their entropy by being strongly entangled.

iii.) Molecular attraction, i.e. the tendency of the system to minimize the potential energy of the van der Waals attraction between all particles.

We shall see that these three tendencies are delicately balanced in the gel, so that slight changes in temperature and slight changes in the density of counterions can make one overwhelm the others, thereby inducing swelling or shrinking. The first two tendencies are entropic and their influence is therefore regulated by temperature. The osmotic pressure will primarily depend on the number of counterions which in turn can be controlled by the acidity of the solvent. Thus we understand the decisive influence of temperature and pH-value on the volume in a qualitative manner.

3. Thermodynamics

According to a simple thermodynamic argument the free enthalpy

$$G = E - TS + pV \tag{3.1}$$

tends to a minimum in equilibrium in a system in which the temperature T is fixed and on whose surface we have the fixed pressure p. The energy of the system is E, its entropy S and V is the volume. Figure 2 shows the system under consideration. Quantities referring to the gel will carry the index I and those referring to the solution bath have the index II. The indices 1, 2 and 3 refer to the solvent, the polymer and the counterions respectively.

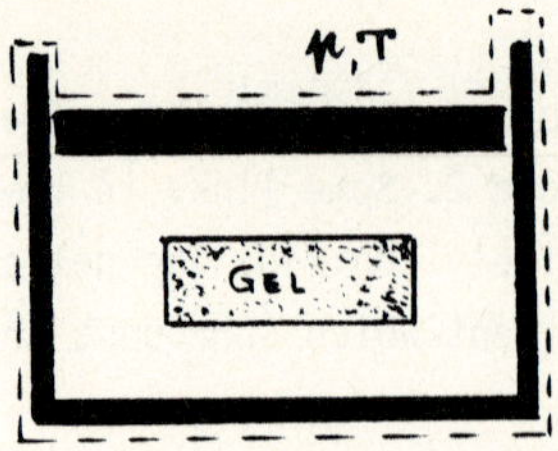

Figure 2: Closed system of solvent and gel

Thus we have

- n_1^I, n_1^{II} - number of solvent molecules in the gel and the solution bath
- n_2 - number of polymer chains
- n_3 - number of counterions

The free enthalpy of a polymeric network in solution has been calculated by Flory [5]. Adapted to the present case of an ionized network with mobile counterions Flory's formula reads

$$\begin{aligned} G^I = {} & n_1^I g_1(T,p) + x n_2 g_2(T,p) + n_3 g_3(T,p) + \\ & + kT \left(n_1^I \ln \frac{n_1^I}{n^I} + n_2 \ln \frac{x n_2}{n^I} + n_3 \ln \frac{n_3}{n^I}\right) + \\ & + e_{12} \frac{n_1^I x n_2}{n^I} + e_{13} \frac{n_1^I n_3}{n^I} + e_{23} \frac{x n_2 n_3}{n^I} \\ & + G_{Network} \end{aligned} \tag{3.2}$$

x is the number of molecule-sized segments of each polymer chain. The first line represents the free enthalpies of the three pure constituents at T and p, while the second and third lines represent entropy of mixing and the heat of mixing respectively. n^I stands for $n_1^I + x n_2 + n_3$ and e_{ij} are the free energies of the bonds between the molecules of the constituents. Always following Flory [5] the free energy of the network is given by

$$G_{Network} = kT \frac{3}{2} n_2 (\alpha^2 - 1 - 2 \ln \alpha) \tag{3.3}$$

where α is the diagonal element of the deformation gradient which in the case under consideration in an isotropic tensor.

The free enthalpy of the solution bath is simply given by

$$G^{II} = n_1^{II} g_1(T,p). \tag{3.4}$$

Of course we must have $n_1^I + n_1^{II} = n_1 = \text{const}$ and therefore $G = G^I + G^{II}$ is a function of the two variables

$$n_1^I \text{ and } \alpha \tag{3.5}$$

provided n_3, the number of ions is fixed. We shall introduce the fraction v_2 of polymersegments in the gel and the fraction R of ionized polymer segments by

$$v_2 = \frac{xn_2}{n^I} \qquad R = \frac{n^3}{xn_2} \quad . \tag{3.6}$$

Assuming the space occupied by the gel to be densely filled by molecules we may reduce the two variables (5) to one, viz. v_2, because we have

$$\alpha = \frac{1}{v_2^{1/3}} \tag{3.7}$$

With a little calculation we may thus write the variable part $\hat{G}$ of the free enthalpy as

$$\hat{G} = \frac{1-(1+R)v_2}{v_2} \ln\left[1-(1+R)v_2\right] + (R+\frac{2}{x}) \ln v_2 -$$

$$- \left[\frac{e_{12}+Re_{13}}{kT}(1+R) - \frac{Re_{23}}{kT}\right] v_2 + \frac{3}{2xv_2^{2/3}} \tag{3.8}$$

We proceed to discuss this formula. Note that v_2 lies between zero and one. $v_2 \lesssim 1$ corresponds to the collapsed gel while $v_2 \gtrsim 0$ corresponds to the swollen gel.

4. The three Contributions to the Free Enthalpy

There are three terms in equation (3.8) which we denote by O, A and N as follows

$$O = \frac{1-(1+R)v_2}{v_2} \ln\left[1-(1+R)v_2\right] + (R+\frac{2}{x}) \ln v_2 \tag{4.1}$$

$$A = - \left[\frac{e_{12}+Re_{23}}{kT}(1+R) - \frac{Re_{23}}{kT}\right] v_2 \tag{4.2}$$

$$N = \frac{3}{2x} \frac{1}{v_2^{2/3}} \quad . \tag{4.3}$$

Figure 3 shows these three contributions as functions of v_2 as well as the sums $N + O$ and $\hat{G} = N + O + A$. x has been chosen as 100 and $e_{12} = e_{13} = e_{23} = 2kT$ in drawing the curves.

O represents the osmotic pressure which is due to the tendency of the counterions to expand the network in order to approach a uniform distribution. Note that O has a minimum for small values of v_2, i.e. in the swollen state. N represents the contractive tendency of the networks and accordingly it is minimal for $v_2 = 1$, i.e. in the collapsed state. These two opposing tendencies taken together produce a minimum at an intermediate value of v_2 as seen from the curve N + O. The expression A in (2) represents the van der Waals potential which, naturally, is minimal in the collapsed state. Figure 3 also shows that all three terms combined produce a free enthalpy with two minima of which we expect the system to assume the lower one.

Comparison of the two-diagrams of Figure 3 shows that a slight shift in the fraction R of ionized polymer segments will reverse the relative height of the two minima and thus lead to a drastic change of volume. For the larger value of R the gel will be swollen and for the smaller value of R it will be collapsed.

In the present theory R is simply a parameter. But it is reasonable to expect that R can be controlled by the acidity of the solvent, so that for a stronger acid there will be less ionisation of the network.This would conform to the observed phenomenon that is illustrated in Figure 1b.

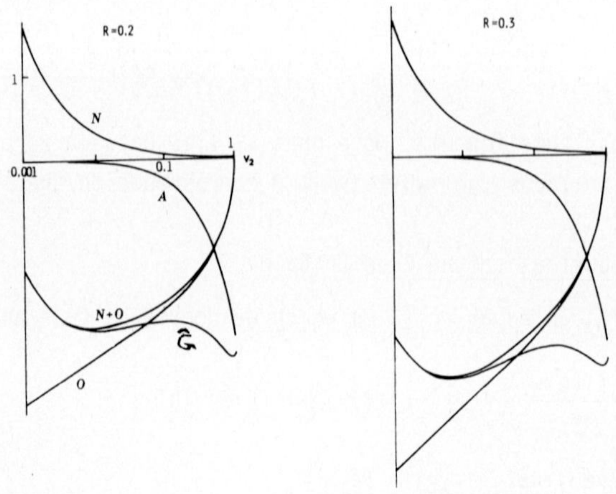

Figure 3. Contributions to the Free Enthalpy and the Free Enthalpy itself

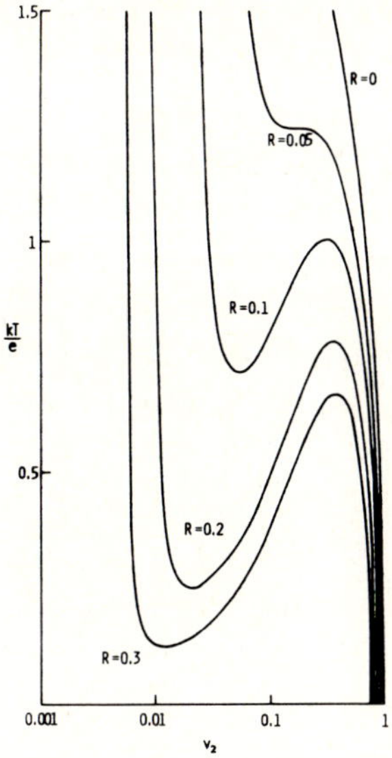

Figure 4. (T,v_2)-curves for several values of R

5. The Effect of Temperature

We return to equation (3.8) which gives the free enthalpy $\hat{G}$ as a function of v_2. $\hat{G}$ must be minimal for equilibrium. In order to find the equilibrium value v_2 as a function of temperature we set the derivative of $\hat{G}$ with respect to v_2 equal to zero and obtain

$$\frac{kT}{e} = - \frac{(1+R+R^2)\, v_2{}^2}{\ln\left[1-(1+R)v_2\right] + (1 - \frac{2}{x})\, v_2 + \frac{1}{x}\, v_2{}^{1/3}} \tag{5.1}$$

where we have set $e_{12} = e_{13} = e_{23} = e$. Figure 4 shows T as a function of v_2 for different values of R as predicted by equation (1). We conclude that in a certain range of values of temperature there are three possible equilibrium values of v_2, if R is small enough.

In order to properly understand the significance of the non-monotone (T,v_2)-curves of Figure 4 let us concentrate on the one for R = 0.2. We draw $\hat{G}$ for that value of R and for different values of T and obtain the curves of Figure 5. The abscissae of the extrema of those curves for a given T are the values of v_2 which satisfy the equilibrium condition(1). Therefore we conclude that only the right and left branches of the non-monotone (T,v_2)-curves actually correspond to equilibria, the central branch represents maxima of $\hat{G}$. The transition from the swollen state to the collapsed state upon a decrease of temperature will occur when the right minimum of $\hat{G}$ becomes lower than the left minimum. This is the case for $\frac{kT}{e} \approx 0.61$. That value has been marked in Figure 6 by the horizontal line.

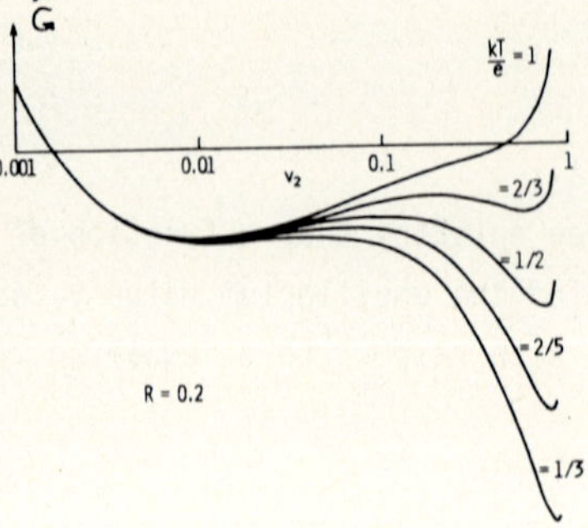

Figure 5. Free Enthalpies for R = 0.2 and different values of T

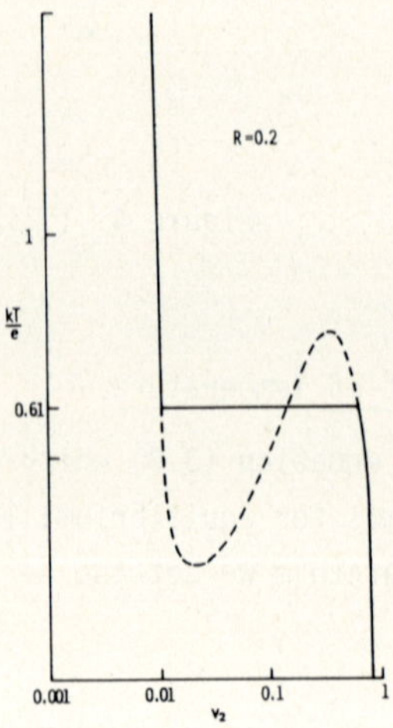

Figure 6. (T,v_2)-curve for R = 0.2 with transition at $\frac{kT}{e} \approx 0.61$

6. Possible Extensions of the Theory

In [4] the present theory has been explained in a slightly more general context. Indeed the gel was not only subject to an isotropic pressure in that paper but in addition to uniaxial and biaxial loads. The results are just as definite as the present ones and can easily be explained suggestively.

A possible refinement of this theory can occur in two directions:

i.) The chemical interaction of the solvent and the ionizing gel should be taken into account. Thus the degree of ionization may be made variable and controllable by the solvent.

ii.) The electrolytic nature of the gel should be studied in detail. This involves a reformulation of the equilibrium conditions in terms of electrochemical potentials.

Work along these lines is in progress. What results have been achieved so far confirm the view that the theory presented in this paper correctly describes the leading phenomena.

References

[1] Tanaka, T., Phys. Rev. Letters 40, 820 (1978)
[2] Tanaka, T., Fillmore, D., Sun, S.T., Nishio, I., Swislow, G., Shah, A. Phys. Rev. Letters 45 , 1636 (1980)
[3] Tanaka, T., Scientific American, January (1981)
[4] Morro,A., Müller, I. Rheologica Acta 27, 44 (1988)
[5] Flory, P.J. Principles of Polymer Science. Cornell University Press, Ithaca and London, 12th edition

ADIABATIC PHASE CHANGES, FAST AND SLOW

Philip A. Thompson and Gerd E.A. Meier
Max-Plank-Institute fur Strömungsforschung
Bunsenstr. 10
D-34 Göttingen
Federal Republic of Germany

ABSTRACT

Many phase changes proceed spontaneously from a metastable state, without significant heat transfert. Diverse examples are given, including vapor explosion, liquefaction shocks, compression to a state near the thermodynamic critical point, shock splitting, liquid-evaporation waves and others. The phase changes discussed are limited to liquid-vapor systems. Homogeneous and heterogeneous nucleation are briefly discussed.

1- INTRODUCTION

Phase changes are a part of everyday experience. Boiling water for tea or coffee, snow or rainfall, vaporization of gasoline in an engine and ice skating are familiar examples. Such phase changes have practical importance, yet still pose unresolved scientific and technical problems, such as the clean combustion of automotive fuels. They can also create scenes of exquisite beauty, such as rainbows, stratospheric clouds, opals, diamonds and snowflakes.

The purpose of this article is to offer an elementary description of a variety of phase-change phenomena in liquid-vapor systems (e.g., water and steam, liquid gasoline and its vapor) especially for phase changes which are not driven by pressure changes or the collapse of a metastable state. We refer to these phase changes, in which heat transfer plays an insignificant role, as adiabatic. An example is the discharge of a CO_2 fire extinguisher, in which the expanding gas is sufficiently coded to produce considerable quantities of solid CO_2 , sometimes called "dry ice". Adiabatic phase changes are common, especially in rapid and/or large-scale events.

It is not possible to give a comprehensive description of this field in the space available here. The reader may therefore wish to regard this article as a set of references complemented by a picture gallery ! Additional references can be found in the papers cited here.

2 - ADIABATIC SYSTEMS

Let us begin with the set of vapor-pressure curves shown in Figure 1. The vapor-pressure curve σ, sometimes called the saturation boundary, divides the region strictly into two equilibrium phases, liquid and vapor (for simplicity, no solid phases are shown). Other boundaries shown include the spinodal curves LS and VS, which limit almost absolutely the extent of possible excursions into adjacent metastable regions. A first example is the constant-pressure heating of a liquid shown in Figure 1a, starting as a subcooled liquid at state 0. As the temperature is increased, the state reaches point 1 on the saturation boundary. In a truly equilibrium World, the liquid would, at this point, begin to boil : in the real world, the liquid will probably increase further in temperature, but very slightly, before boiling begins, i.e., the state would be slightly supersaturated. In unusual circumstances, perhaps using a very clean glass flask for the liquid container, the temperature may greatly exceed the equilibrium boiling temperature T_1. When boiling does finally occur, it does so with extreme violence, a phenomenon sometimes observed in chemistry laboratories. In the most extreme case, the state almost reaches the spinodal limit LS at temperature T_2, which may be more than 200°C above the equilibrium boiling point, depending, of course, on the liquid being used in the experiment. In this case, the sudden and violent evaporation which takes place is called a vapor explosion or explosive boiling. Such an event usually requires a special set of circumstances : See Skripov (1974), Shepherd & Sturtevant (1983) and Frost & Cicciarelli (1987).

It should be remarked that explosive boiling is produced by heating of the liquid : once the explosion is initiated, however, there is very little time available for heat transfer, so that the actual phase change can usually be considered to be adiabatic.

Vapor-explosion accidents, although unusual, lead to a considerable loss of human life.

Another means of evaporating water employs isothermal decompression, shown in Figure 1b, leads to a similar sequence of states 0,1,2. The extreme state (2) at the spinodal limit is a state of hydrostatic tension. The hydrostatic tension found by Briggs (1950) in water was well short of the spinodal, however.

Figure 1c shows the isothermal compression of a vapor, leading to liquefaction. The Wilson line W shown represents the maximum pressure at which the onset of condensation is observed in very rapid processes. To maintain an isothermal state, a very slow process is required and the effective onset of condensation occurs just above the vapor-pressure curve. The vapor states 2 and 3 are not realized in this case.

Figures 1d and 1e represent two distinct cases of vapor condensation. Case d shows condensation of a "regular" substance, such as water, by isentropic decompression (expansion).

This process occurs, for example, in the interior of steam turbines and in cloud formation. Starting from an initial dry-vapor state 0, the expansion proceeds past the equilibrium saturation boundary at 1 without condensation occurring at or near the Wilson line W.
There is no evidence that the expansion of the vapor goes much further than this, i.e., the vapor spinodal is not approached and thus becomes a somewhat academic limit.

Case e shows condensation of a "retrograde" substance, such as iso-octane, by isentropic compression. Again, the onset of condensation occurs near the Wilson line at point 2. State 3 on the vapor spinodal is not approached, except possibly in the case of states near the thermodynamic critical point.

The distinction between regular and retrograde fluids is illustrated in Figure 2 using a temperature-entropy diagram. When the vapor saturation boundary (right boundary) is inclined to the right, as in the case of octane, the fluid is said to be retrograde ; otherwise, it is said to be regular, as in the case of ethene. A vertical line on the temperature-entropy diagram corresponds to an isentropic process. It can be seen from the diagrams that regular fluids condense on isentropic expansion and retrograde fluids condense on isentropic compression, consistent with Figures 1d and 1e. This behavior is discussed at length in Thompson, Carofano & Kim (1986). It is easy to show that retrograde behavior is associated with large molar heat capacity (many vibrational degrees of freedom). It is convenient to define a nondimensional characteristic heat capacity $\tilde{C}_V$,

$$\tilde{C} \equiv C_V^0 (T_C) / R \tag{1}$$

where $C_V{}^o$ is the ideal-gas heat capacity, T_c is the critical temperature and R is the gas constant. It is found that the behavior will be locally retrograde for $\tilde{C} > 11.2$, a value that corresponds roughly to butane or sulfur hexafluoride, i.e., these substances are marginally retrograde.

The distinction between regular and retrograde fluids was already known to van der Waals (1908) and Planck (1903). For our purposes, it is important to note tat the adiabatic phase changes observed in these two catégories of fluidd are different in kind.

In addition to phase changes in regular and retrograde fluids, other two-phase systems should be mentioned. For example, the graphite-diamond system is closely analogous to the retrograde liquid-vapor system, as discussed by Thompson, Kim & Carofano (1986), Thompson,

Chaves, et al. (1987) ; see also Mc Queen & Marsh (1968) and Gust & Young (1979). The system of a gas in equilibrium with a liquid and soluble in that liquid (e.g., a bottle of mineral water with dissolved carbon dioxide) is also analogous to the retrograde liquid-vapor system. Adibatic ice-liquid systems are also interesting, for example, in connection with ice skating. Finally, the behavior of near-critical liquid-vapor systems is a special category which leads to surprising phenomena related to phase changes : see Borisov, et al. (1983), Kutateladze, et al. (1987) and Zauner (1988).

3 - SUPERSATURATION, TIME SCALES AND NUCLEATION

For present purposes, the formation of a new phase involves either the condensation of vapor into the liquid phase or the evaporation of liquid into the vapor phase. The phase transtion requires a series of steps : the old phase must first be brought to saturation (i.e., to the vapor-pressure curve) and then into supersaturation, the amount of supersaturation depending on the rapidity of the process. The old phase is now in a metastable state favorable to the formation of nuclei (microscopic 'seeds') or to the activation of existing nuclei (e.g., atmospheric particles or ting bubbles in a liquid). Nucleation and growth of the new phase may then proceed. As a specific and relevant example, the condensation of vapor to liquid illustrated in Figures 1d and 1e proceeds from state 0 (dry, superheated vapor) to state 1 (saturated vapor) to state 2 (metastable vapor), where nucleation of submicroscopic liquid droplets begins, followed by droplet growth and substantial conversion of vapor to liquid. In the case of conversion by isentropic expansion-illustrated in Figure 1d, the end temperature may well be low enough that solid particles (ice) are formed, rather than liquid.

The supersaturation S is usually defined for vapor-condensation experiments, as discussed above, to be

$$S \equiv \frac{P}{P_{\sigma}(T)} \geq 1 \tag{2}$$

The value of S varies greatly. At the critical point, there is fairly good evidence that the vapor will not supersaturate, that is, that $S = 1$, the minimum possible value. More typical values are 1.02 for cumulus cloud formation and 50 for shock-tube expansion experiments. Critical supersaturation is roughly equivalent to the Wilson line : it corresponds to a state at which further supersaturation would lead to an enormous increase in the nucleation rate and thus approximately defines the onset of condensation.

In the case of the supersaturation of a liquid, for example in a vapor-explosion experiment, it is customary to characterize the supersaturation by the excess of temperature above the normal saturation temperature, i.e. by $T - T_\sigma(P)$, called the superheat. It would, of course, be possible to retain the definition (2) for liquids. One difficulty in dynamic experiments is the measurement of local pressure.

For recent work in the onset of nucleation, the reader is referred to Bier, et al. (1977), Bier et al. (1984), Dobbins (1983), Peters (1987), Wegener (1987), Wegener & Mirabel (1987) and Wegener & Wu (1977).

The preceding discussion and the references given deal primarily with homogeneous nucleation, which develops from spontaneous molecular attraction and does not depend on foreign particles or foreign surfaces. In general, there seem to be basically three kinds of nucleation :

1. Homogeneous nucleation, as described above. It is usually associated with large departures from equilibrium (large supersaturations) and often proceeds with great rapidity. In some cases, a clean system is required.

2. Heterogeneous nucleation on atmospheric particulates, aerosols and foreign surfaces. In effect, the required nuclei are already present or have been artificially introduced, as in cloud seeding. Surface defects, such as cracks and pits may enhance nucleation. Atmospheric precipitation (rain, snow) is almost invariably associated with heterogeneous nucleation.

3. Preexisting microbubbles. In many liquids, including water, microbubbles with diameters in the order of one micron are present in large concentrations. If the pressure on the liquid is then reduced, as normally required for adiabatic evaporation, the microbubbles will grow naturally. Such conditions are favorable to violent cavitation. In this case, there is no nucleation : the bubbles are already present. See Rein (1987).

An example of heterogeneous nucleation is shown in Figure 3. In this unusual case, the actual lead particle which served as the seed has been preserved. An example of vapor condensation originating from homogeneous nucleation is shown in Figure 4. The liquefaction shockwave shown is strong enough to completely liquify the vapor ahead. See also : Dettleff, Meier, et al. (1982) and Dettleff, Thompson, et al. (1979).

Most of the processes of interest here are short-time events. In the liquefaction shock shown in Figure 4, for example, the estimated time for a fluid particle to pass through the shockfront is 1/2 10^{-5} seconds. In this very short time, very large supersaturation is built up, leading to very rapid nucleation and condensation. A very large departure from equilibrium has been resolved by a correspondingly drastic process to restore equilibrium. A similar description of the vapor explosion described in section 2. A reasonable attempt to generalize these observations might run as follows : "The greater the departure from equilibrium, the more violent the readjustment to a new equilibrium". The commonly encounterd departures from equilibrium, especially those involving metastable states, must be incorporated into mathematical models of phase transitions.

Not all phenomena are as brissant as the liquefaction shock described. The condensation of moist air to form a cloud is a comparatively liesurely process, governed by modest upward convection velocities and leading to quite small supersaturations, as already mentioned.

4 - PRODUCTION OF A METASTABLE STATE AND SUBSEQUENT PHASE CHANGE : EXAMPLES

A considerable variety of adiabatic phase changes is listed in Table 1, which is arranged in chronological order (the writers would be very grateful for documented corrections). The bottom part of this table lists phenomena which are anticipated theoretically but not yet demonstrated. In the following, several of the phenomena will be described, including one not listed in Table 1.

It is useful to categorize the phase changes according to their driving agents (e.g., decompression) as was done in Figure 1. The driving agent produces a metastable state, followed by nucleation and phase transition. The categories and their corresponding phase transitions are :

1. Heating (vapor explosion).

2. Compression (liquefaction shock ; shock compression to a near-critical state ; shock splitting with phase change).

3. Decompression or expansion (liquid-evaporation wave ; cavition, with or without extreme negative pressures ; moisture "shocks" in Laval nozzles or turbines ; champagne effect ; rain and snow ; expansion from a critical state, e.g., leading to a rarefaction shock, or a steady expansion flow at the critical state ; moisture-evaporation rarefaction shock ; single-phase rarefaction shock).

This makes a total of 12 phase changes (including the single-phase rarefaction shock, which is phase-transition related). These phase transitions are now described briefly in turn.

The vapor explosion or explosive boiling (skripov) has already in section 2. The extreme superheats that are near the spinodal limit usually require that the test liquid be surrounded by a "host" liquid of lower vapor pressure, in order to avoid heterogeneous nucleation.

The liquefaction shock, illustrated in Figure 4 can be either partial or complete, depending on whether all of the vapor is converted to liquid. A characteristic heat capacity $\tilde{C} > 24$ is required to produce a liquefaction shock. This phenomenon is described in detail by Thompson, et al. (1986) and earlier papers cited there.

Shock compression to a near-critical state represents a special form of liquefaction shock, in which the state downstream of the shock front is supposed to be near the thermodynamic critical point (this "dynamic" critical state may, however, be shifted in pressure and temperature). A preliminary experiment is reported by Thompson, Kim, Yoon & Chan (1988), who used the test fluid n-C_6F_{14}. Photographs of the shockfront shown in Figure 5 show some peculiar features, for example, the very small divergence of the jet emerging from the tube. This suggests a very small soundspeed in the jet fluid, corresponding to the slow rate of expansion. The appearence of large vortex rings is also of interest. Theoretical suggestions for this experiment have been made by Zel'dovich (1981).

Shock splitting was one of the possibilities investigated by Bethe (1942), along with the single-phase rarefaction shock. The existence of shock splitting has now been amply confirmed, see Thompson et al. (1987). Theoretical studies have been made by Cramer (1988), Fowles & Houwing (1984), Menikoff & Plohr (1988) and Zel'dovich & Raizer (1967). Illustrations of experimental results are shown in Figures 6 and 7, which clearly show the devolution of a single liquefaction shock into a vapor-phase shockfront and a condensation discontinuity. Figure 8 shows the system of calculated adiabats and the corresponding states of the fluid in one-dimensional flow.The various states are as follows :

0 Initial vapor state agead of the advancing shockfront system.

1 State of critical supersaturation, corresponding to the vapor-phase state at the triple point.

2 Equilibrium two-phase state at the triple point, corresponding to the state behind the liquefaction shock.

CA Possible two-phase equilibrium states.

DA Possible supersaturated vapor states.

Liquid-evaporation waves are found in metastable liquids following sudden depressurization of a vessel, as found in the experiments of Chaves (1980, 1984) and described by Thompson, et al. (1987). It appears that this phenomenon was first observed by Grolmes & Fauske (1974).

Cavitation is an enormous field to which we can offer little illumination. Theoretically, cavitation waves should split and Table 1 shows a reference which appears to offer experimental confirmation.

The moisture shock refers to the sudden condensation of metastable vapor in high-speed flow, for example in the flow of moist air through a Laval nozzle. It represents a practical problem in turbines and wind tunnels (Wegener & Mack 1958). Its discovery goes back to an "unsuccessful" experiment of Ludwig Prandl in Göttingen (Prandtl 1940) ; Hermann (1942). It has been estimated that moisture shocks can result in an increase of 1 % in the fuel consumption of a jet airplane (David Hobbs, private communication 1982). The original photograph from Prandtl's laboratory is shown in Figure 9. The visible X-shock has been induced by a moisture shock which is not visible. The first complete theory of the moisture shock is due to Oswatitsch (1942).

The term Champagne effect derives from the spouting of a bottle of too-warm Champagne. In current usage, it refers to the uncontrolled (and occasionaly catastrophic) release of dissolved gas from a host liquid, which is typically water. It is a technical problem in the underground storage of compressed air (Sneck, et al. 1985) and in reactor emergency core cooling (Simoneau 1981). The application to be described here is the natural catastrophe which occurred on the 21 of August 1986 at Lake Nyos, in Cameroon, West Africa. The brief account given here is largely based on the paper of Dling, Clard, et al. (1987). On the day of the catastrophe, the lake was saturated (or possibly supersaturated) with carbon dioxide of magmatic origin. The physical basis for the disaster is in Henry's Law, which states that the solubility of a gas in a liquid is proportional to pressure. In the case of Lake Nyos, this means that the equilibrium concentration of CO_2 would increase approximately linearly with depth. The depth of the Lake is about 200 m, based on a hydrostatic pressure gradient of about one atmosphere per 10 meters change in depth. Thus, the concentration at depth would be about twenty times the concentration at the surface. The total dissolved CO_2 in Lake Nyos prior to the disaster is estimated to be equivalent to a CO_2 cloud of about 1.5 cubic kilometers at one atmosphere.

A liquid column or a lake containing large amounts of dissolved gas can be unstable. Consider a small packet of liquid which is disturbed and carried upward by convection to a point where the local hydrostatic pressure is smaller : then the small bubbles in the packet will expand, the packet acquires buoyancy and the upwfard motion is accelerated. The packet then expands further,and so on. In the case of Lake Nyos, a plansible disturbance for the initiation of this chain could be a small landslide.

The instability, once started, produced spectacular effects, including a two-phase fountain that produced the high wash shown in Figure 10. The enormous cloud of released CO_2 was more dense than the surrounding air (m the ratio 44/29) and rapidly overflowed the volcanic basin of Lake Nyos into the low-lying areas. 1 700 human beings and 3 000 cattle died of $C0_2$ asphyxiation.

Expansion from a critical state has produced rarefaction shockwaves, as reported by Borisov, et al. (1983). Such shockwaves are possible because the fundamental derivative T $\left[\text{corresponding to} \left(\frac{\partial^2 P}{\partial V^2}\right)_S\right]$ becomes negative as C_V increasing in value according to the critical power law. The expansion experiments of Zauner (1988) are quasisteady flow experiments and show unusual effects whose interpretation is not yet clear. The authors plan to pursue experiments of this kind, with steady flow, in the near future.

Some results from the near-critical shock-compression experiments already referred to are shown in Figure 11. The soundspeed was measured by a method based on Skews construction (Skews 1967 ; Chen, et al. 1986), leading to soundspeeds of about 10 m/s in n-C_6F_{14}. This value is roughly an order of magnitude of previous near-critical values (e.g. Garland & Williams 1974). The "settling time" calculations for this preliminary experiment with C_6F_{14} were based on a method suggested by Moldover & Gammon (1983).

The moisture-evaporation rarefaction shock converts a liquid-vapor mixture into a superheated vapor across a shockfront, using a retrograde fluid (Thompson, Carofano & Kim 1986). In the experiments described, this shock had the unusual property of both sonic inflow and sonic outflowa doubled Chapman-Jouguet condition.

The single-phase rarefaction shock is the classical "impossible" shock. It has been jound in many materials (e.g., quartz, iron), but not in a liquid or a vapor. There are, however, ample theoretical studies which show the plausibility of their existence (Cramer & Kluwick 1984).

5 - CLOSING REMARKS

Sophisticated mathematical models of phase changes based on capillarity (e.g., Slemrod 1983 and 1984) have not yet reached the stage of realistic description of actual phase changes.

When that stage is reached, the rewards will be great. In the meantime, the analytical modeling needs improvement, especially in the equations of state-primarily in the near-critical region [The equations of state used in the work reported here were primarily those of Abbott (1973) and Hobbs (1983)].

For those with further interest in phase changes, a useful summary is given by Meier (1987).

There are numerous secondary phenomena associated with phase changes which have not been mentioned. One of these is shockfront instability, visible in an incipient form in Figure 6 as a "gof-ball" pattern. These instabilities can grow and diminish, depending on the strength of the shock. Their limits have been described, more or less successfully, according to the model of Kontorovich (1957) : see Fowles & Houwing (1984) and Thompson, et al. (1986).

The authors wish to thank the students and colleagues who have participated in past research. We also thand those whose material we have used. The senior author gratefully acknowledhes the support of the National Science Foundation (USA) under grant MSM 8412849 and the assistance of the Max-Planck-Gesellschaft.

TABLE 1 : ADIABATIC PHASE TRANSITIONS IN LIQUID-VAPOR SYSTEMS.
(Partial List)

TRANSITION PHENOMENON	AUTHOR	YEA
Rain and Snow	Old Testament	
Turbine Expansion/ Condensation	Heron of Alexandria	120 BC
Vapor Explosion	Dufour	1861
Supersaturation of a Vapor Jet	Helmholtz	1886
Cavitation in Water	Reynolds	1894
Condensation-Induced Shock ('moisture shock')	Prandtl/Wieselsberger	1935
Prediction of Vapor-Phase Rarefaction Shock	Zel'dovich	1946
Negative Liquid Pressure	Briggs	1950
Liquid-Evaporation Wave	Grolmes/Fauske	1974
Cavitation - Wave Splitting (?)	Yamaguchi/Ichikawa	1977
Complete and Partial Liquefaction Shocks	Dettleff/Thompson/Meier	1979
Rarefaction Shock from a Critical State	Borisov et al.	1983
Shock Splitting	Speckmann/Thompson/Kim	1983
Chapman-Jouguet Liquid-Evaporation Wave	Chaves	1984
Mixture-Evaporation Rarefaction Shock	Thompson/Carofano/Kim	1986
Shock to a Thermodynamic Critical State	Thompson, et al. ?	1988

Sonic Flow at the Thermodynamic Critical Point ?

Shock Splitting in a Single-Phase Fluid ?

Complete Evaporation Shock ?

Single-Phase Rarefaction Shock ?

Approach to the Vapor Spinodal ?

REFRENCES

ABBOTT, M.M. 1973 Cubic equations of state. *AIChE. J.* **19**, 596-601.

BETHE, H.A. 1942 The theory of shock waves for an arbitrary equation of state. *Rep.* 545, p. 57.
Office of Scientific Research and Development, Washington.

Bier, K., Ehrler, F. Kissau, G., Lippig, F. & Schorsch, R. 1977
Homogene Spontankondensation in expandierenden Dampfstrahlen des Kältemittels R 22 bei hohen normierdten Drüken. Forsch. *Ing. Wes.* **43**, 165-175 (V.D.I.).

Bier, K., Ehrler, F. & Theis, G. 1984 Comparison of spontaneous condensation in supersaturated nozzle flow of different refrigerants. ORC-HP *Technology* , VDI Bericht 539, pp. 749-768. VDI Verlag, Düsseldorf.

Borisov, A.A., Borisov, Al. A. & Kutateladze, S.S. 1983 Rarefaction shock wave near the critical liquid-vapour point. *J. Fluid Mech.* **126**, 59-73.

CHAVES, H. 1980. Verdampfungswellen in retrograden Fluessigkeiten. Diplomarbeit, Georg-August-Universitaet, Goettingen.

CHAVES, H. 1984 Phasenubergaenge und Wellen bei der Entspannung von Fluiden hoher spezifischer Waerme. Dissertation, Georg-August Universitaet, Goettingen.

Chen, G., Thompson, P.A. & Bursik, J.W. 1986 Soundspeed measurements in vapor-liquid mixtures behind shock waves. *Experiments in Fluids.* 4, 279-282.

CRAMER, M.S. & Kluwich, A. 1984 On the propagation of waves exhibiting both positive and negative nonlinearity. *J. Fluid Mech.* **142**, 9-37.

CRAMER, M.S. Shock splitting in single-phase gases. To appear, J. Fluid Mech.

DETTLEFF, G., Meier, G.E.A., SPECKMANN, H.-D., THOMPSON, P.A. & YOON, C. 1982
Experiments in shock liquefaction. *In Proc. 13th Intl Symp. on Shock Tubes and Waves* (ed. C.E. Trainor & J.G. Hall), pp. 616-623. State University of New York Press, Albany.

DETTLEFF, G., THOMPSON, P.A., MEIER, G.E.A. & SPECKMANN, H.-D. 1979 An experimental study of liquefaction shock waves. *J. Fluid Mech.* **95**, 279-304.

DOBBINS, R.A. 1983 A theory of the Wilson line for steam at low pressures. *Trans.* ASME I : *J. Fluids Engng* **105**, 414-422.

FOWLES, G.R. & HOUWING, A.F.P. 1984 Instability of shock and detonation waves. *Phys. Fluids* **27**, 1982-1990.

FROST, DAVID L. & CICCIARELLI, G. 1987 Dynamics of explosive interactions between multiple drops of tin and water. Proc. 11th Int. Colloquin on Dynamics of Explosions and Reactive Systems, Warsaw.

GARLAND; C.W. & WILLIAMS, R.D. 1974 Low-frequency sound velocity near the critical point of xenon. *Phys. Rev.* A. 10, 1328-1332.

GUST, W.H.&YOUNG, D.A. 1979 *High Pressure Science and Technology* (ed. K. D. Timmerhaus & M. S. Barber), vol. I, pp. 944-852. Plenum.

HERMANN, R. 1942 Der Kondensationsstoss in Uberschall-Windkanal düsen. Luftfahrtforschung 19, 201 - 209.

HOBBS, D.E. 1983 Avirial equation of state utilizing the principle of corresponding states Dissertation, Rensselaer Polytechnic Institute.

KLING GEORGE W., MICHAEL A.CLARK, HARRY R. COMPTON, JOSEPH D. DEVINE, WILLIAM C. EVANS, ALAN M. HUMPHREY, EDWARD J. KOENIGSBERG, JOHN P. LOCKWOOD, MICHELE L. TUTTLE, GLENN. WAGNER 1987

The 1986 Lake Nyos gas disaster in Cameroon, West Africa. Science 236, 169-175.

KONTOROVICH, V.M. 1957 Concerning the stability of shock waves. *Sov. Phys : Tech. Phys.* 6, 1179-1181.

KUTATELADZE, S.S., NAKORYAKOV, V.E. & Borisov, A.A. 1987 Rarefaction waves in liquid and gas-liquid media. In *Annual Rev. Fluid Mech.* 19, 577-600

LANDAU, L.D. & Lifshitz, E. M. 1959 *Fluid Mechanics*, p. 496. Pergamon.

McQUEEN, R. G. & MARSH, S. P. 1968 Hugoniots of graphites of various initial densities and the equation of state of carbon. In *Behavior of Dense Media under High Dynamic Pressures*, pp. 207-216. Gordon and Breach.

MEIER, G.E.A. 1987 Zur Realgasdynamik der Fluide hoher spezifischer Wärme. Habilitationsschrift (Physik), Georg-August-Universität, Göttingen.

MENIKOFF, RALPH& PLOHR, BRADLEY 1988 Riemann problem for fluid flow of real materials. Los Alamos National Laboratory Report LA-UR-88-49.

MOLDOVER, M.R. & GAMMON, R.W. 1983 Conceptual design of low-gravity experiments on phase transition and critical phenomena in fluids. NASA Contractor's Report CR 174637.

OSWATISCH, K. 1942 Kondensationserscheinungen in Uberschalldüsen. ZAMM 22, 1 - 14.

PETERS, F. 1987 Condensation of supersaturated water vapor at low temperatures in a shock tube. J. Phys. Chem. 91, 2487 - 2489.

PLANCK, M. 1903 *Treatise on Thermodynamics* , pp. 150-152. Longmans-Green

PRANDTL, LUDWIG 1940 Remarks. Atti del V Convegno Volta p 196-197 ; 558. Reale Accademia D'Italia, Roma.

REIN, MARTIN 1987 Numerische Untersuchung der Dynamik heterogener StoBKavitation. Dissertation, Georg August Universitaet, Goettingen.

SHEPHERD, J. E. & STURTEVANT, B. 1983 Rapid evaporation at the superheat limit. J. *Fluid Mech.* 121, 379-402.

SIMONEAU, ROBERT J. 1981Depressurization and two-phase flow of water containing high levels of dissolved nitrogen gas. NASA Technical Paper 1839, Lewis Research Center, Cleveland, Ohio.

SKEWS, B.W.1967 The perturbed region behind a diffracting shock wave. *J. Fluid Mech.* 29, 705-719.

SKRIGOV, V.P. 1974 Metastable Liquids. John Wiley, New-York.

SLEMROD, M. 1983 Admissibility criteria for propagating phas boundaries in a van der Waals fluid. *Arch. Rat. Mech. Anal.* **81**, 301-315.

SLEMROD, M. 1984 Dynamic phase transitions in a van der Waals fluid. *J. Diffl Equat.* **52**, 1-23.

SNECK, H.JAMES, Thompson, P.A., Hand, B.E., Meyer, B.R. & Ping, Chen 1985 Modeling of the champagne effect, in Fundamental Aspects of Gas-Liquid Flows Amer. Soc. Mech. Engrs. FED-29.

STODOLA, A. 1924 Dampf-und Gasturbinen Julius Springer Verlag, 6th Ed. (p1).

THOMPSON, P.A., Carofano, G.C. & Kim, Y.-G. 1986 Shock waves and phase changes in a large-heat-capacity fluid emerging from a tube. *J. Fluid Mech.* **166**, 57-92.

THOMPSON, P.A., CHAVES, H., Meier, G.E.A., Kim, Y.-G & Speckmann, H.-D. 1987 Wave splitting in a fluid of large heat capacity. J. *Fluid Mech.* **185**, 385-414.

THOMPSON, P.A., Kim, Y.-G., Yoon, C.J. & Chan, Y. 1988 Nonequilibrium, near-critical states in shock-tube experiments. *Proc. 16th Intl. Symp. Shock Tubes and Waves*, Aachen (ed. Hans Grönig), pp. 343-349. VCH Verlag, Weinheim, FRG.

VAN DER WAALS, J.D. 1908 *Lehrbuch der Thermodynamik.* Bearbeitet von Ph. Kohnstamm, Maas und Van Suchtelen, Leipzig.

WEGENER, P.P. 1987 Nucleation of nitrogen : Experiment and theory. J. Phys. Chem. 91 2479-2481

WEGENER, P.P. & MACK, L.M. 1958 Condensation in supersonic and hypersonic wind tunnels. In *Advances in Applied Mechanics,* vol. 5 (ed. H. L. Dryden & Th. von Karman), pp. 307-447.Academic.

WEGENER, P.P. & MIRABEL, PHILIPPE 1987 Homogeneous nucleation in supersaturated vapors. Natuwissenschaften 74, 111 - 119.

WEGENER, P.P. & WU, B.J.C. 1977 Gasdynamics dand homogeneous nucleation. *Adv. Colloid Interface Sci.* **7**, 325-417.

ZAUNER, E. 1988 Transkritische Expansionsströmungen. Lecture presented at the GAMM Annual Meeting, Sitzung 60, Wien.

ZEL'DOVICH, YA. B. & RAIZER, YU. P. 1967 *Physics of Shock Waves and High-Temperature Hydrodynamic Phenomena.* Vol. 2 (ed. W.D.Hayes & R.F.Probstein), pp. 750-756. Academic.

ZEL'DOVICH, YA. B., "The reestablishment of the van der Waals critical point in fast processes", Soviet Physics JETP, Vol.53, p.1101, 1981.

ADDITIONAL REFERENCES FOR TABLE 1

VAPOUR EXPLOSION (1861)
Dufour, L. C.R. Acad. Sci. Paris 53, 846.

CAVITATION (1894)
Reynolds, Osborne. Papers on Mechanical and Physical Subjects, Volume 2, 63, 578-587. Cambridge Univ. Press, 1901.

SUPERSATURATION OF WATER VAPOUR
Wilson, C.T.R. Trans. Roy. Soc. A 189, 265-307. (1897)

HELMHOLTZ, R.Von, Ann. Phys. Chem. 27, 508. (1886)

PREDICTION OF VAPOUR-PHASE RAREFACTION SHOCK (1946)
Zel'dovich, Ya. B. Zh.Eksp. Teor. Fiz 4, 363-364.

NEGATIVE LIQUID PRESSURE (1950)
Briggs, L. J. J. Appl. Phys. 21, 721.

LIQUID-EVAPORATION WAVE (1974)
Grolmes, M.A. & Fauske, H.K. Proc. 5th Int. Conf. on Pressure Surges, BHRA, London. (Column separation)

COMPLETE AND PARTIAL LIQUEFACTION SHOCKS (1979)
Dettleff, G., Thompson, P.A., Meier, G.E.A. & Speckmann, H.D. J. Fluid Mech. 126, 59-73.

SHOCK SPLITTING (1983)
Thompson, P.A. & Kim, Y.-g. Phys Fluids 26, 3211-3215.
See also Dettleff, et al. 1982.

CHAPMAN-JOUGUET LIQUD-EVAPORATION WAVE (1984)
Chaves, H. Dissertation, Göttingen. See also Thompson, et al 1987.

MIXTURE-EVAPORATION RAREFACTION SHOCK (1986)
Thompson, P.A., Carofano, G.C. & Kim, Y-G. J. Fluid Mech. 156, 57-92.

CAPTIONS FOR FIGURES

1 - Evaporation of liquid and condensation of vapor.

a) Boiling liquid by constant-pressure heating.
b) Boiling liquid by isothermal decompression.
c) Condensing vapor by isothermal compression.
d) Condensing vapor by isentropic expansion (regular fluid, which condenses on isentropic expansion).
e) Condensing vapor by isentropic compression (retrograde fluid, which condenses on isentropic compression).

Symbols : σ = vapor-pressure curve (saturation boundary) ; LS = Liquid Spinodal curve ;VS = Vapor Spinodal curve ; W = Wilson Line, or curve of critical super = saturation for the vapor.

2 - Temperature-entropy diagrams for various substances with increasing heat capacity C_v. Reduced temperature T = T/T_c . Only the liquid and vapor saturation boundaries are shown. a) Ethene C_2H_4, C_v = 4.03 ; b) Propane C_3H_8, C_v = 9.64 ; c) n-Octane C_8H_{18}, C_v = 36.8 ; d) Pf-methyldecaline $C_{11}F_{20}$, C_v = 73.4.

3 - Heterogeneous nucleation of water ice on a particle of atmospheric lead, visible at the center of the large crystal, catalized by trace amounts' of iodine vapor. Electron micrograph by Roger Cheng and Austin Hogan, Atmospheric Sciences Research Center, Albany, New York.

4 - Homogeneous nucleation in a liquefaction shockfront. Oblique photograph showing vortex rings (nucleation related ?). The shockfront is emerging from the shock-tube test section, traveling from left to right. The dark region of the shockfront is liquid. The test substance is n-C_6F_{14}. Initial conditions (ahead of the shock) are P_0 = 2.78 bars, T_0 = 110°C. Shock Mach number M_0 = 3.12.

5 - Near-critical shockfront structure emerging from the end of the shocktube test section. Test substance n-C_6F_{14}. Initial conditions P_0 = 2.78 bars, T_0 = 110°C. Shock Mach number M0 = 2.95. a) Oblique view. b) Profile view.

6 - Shock splitting in iso-octane. The emergence of a liquefaction shockfront from the shock-tube test section is shown in a oblique view (a) and profile view (b). Near the top and bottom of the shockfront, the shock profile has become rounded and a single phase shockfront has seperated from the main two-phase shockfront, as shown from the refraction patterns. The arrow in (b) indicates the triple point, where splitting occurs. Initial conditions $P_0 = 1.1$ bar, $T_0 = 110°C$. Shock Mach number $M_0 = 2.79$.

7 - Closeup view of shock splitting at the triple point in the fluorocarbon FC-75. Initial conditions $P_0 = O.95$ bars, $T_0 = 110°C$. Plane shock Mach number $M_0 = 2.04$.

8 - a) The system of shock adiabats for shock splitting in FC-75, shown in the entropyvolume plane, based on initial conditions $P_0 = 1.80$ bar, $T_0 = 135°C$, $V_0 = 40.85\ cm^3/g$. EA = equilibrium adiabat, CA = condensation adiabatic, DA = dry adiabat (metastable states), o = saturated vapor boundary, W = Wilson line. The horizontal axis indicates the value of V/V_0, i.e., the normalized specific volume.
b) Sequence of states in one-dimensional flow. The forerunner shock FS and condensation discontinuity CD are shown. For triple-point conditions, $U_{FS} = U_{CD}$.

9 - The X-shock observed in moist air by Ludwing Prandtl (Schlieren photograph by A. Busemann). Flow is from left to right.

10 - Soundspeed and pressure measurements for near-critical states in n - $C_6 F_{14}$. P_2 and c_2 are the reduced pressure and soundspeed, respectively, behind the shockfront. The curve for c_2 is a least-squares third-order polynomial fit to teh data. The curve for P_2 is calculated from the Rankine-Hugoniot equations.

11 - Lake Nyos in Cameroon, August 1986. A fountain of water/CO_2 mixture has washed over the 80 - m rock promontory shown. [Michael Clark, Armed Forces Institute of Pathology, Washington D.C.]

12 - The first steam turbine, reported by Heron of Alexandria (120 BC ?). The expelled steam drives the rotating sphere by a form of jet propulsion. Illustration from Stodola (1924).

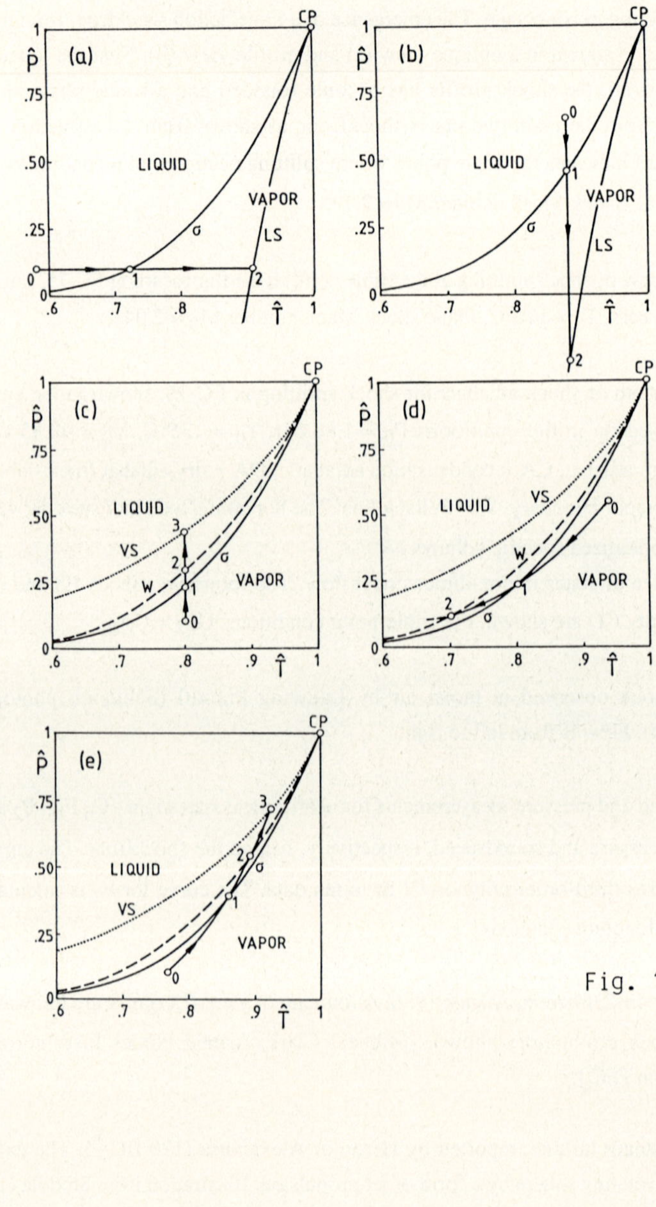

Fig. 1

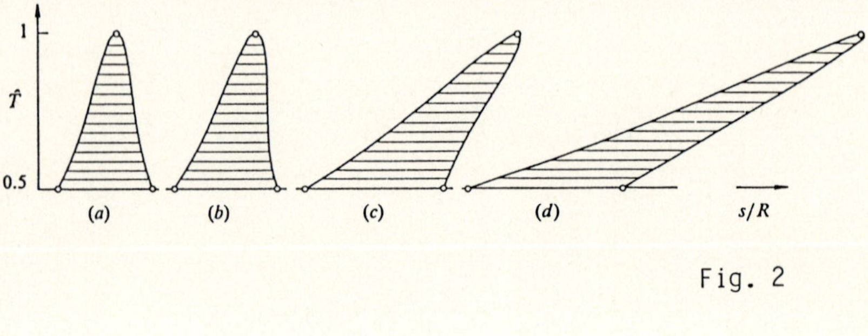

Fig. 2

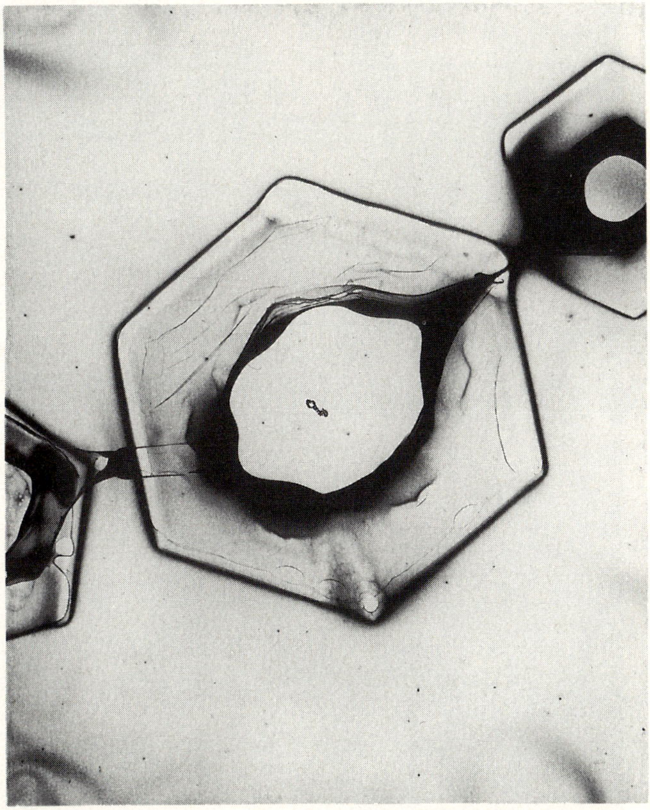

Fig. 3

Fig. 4

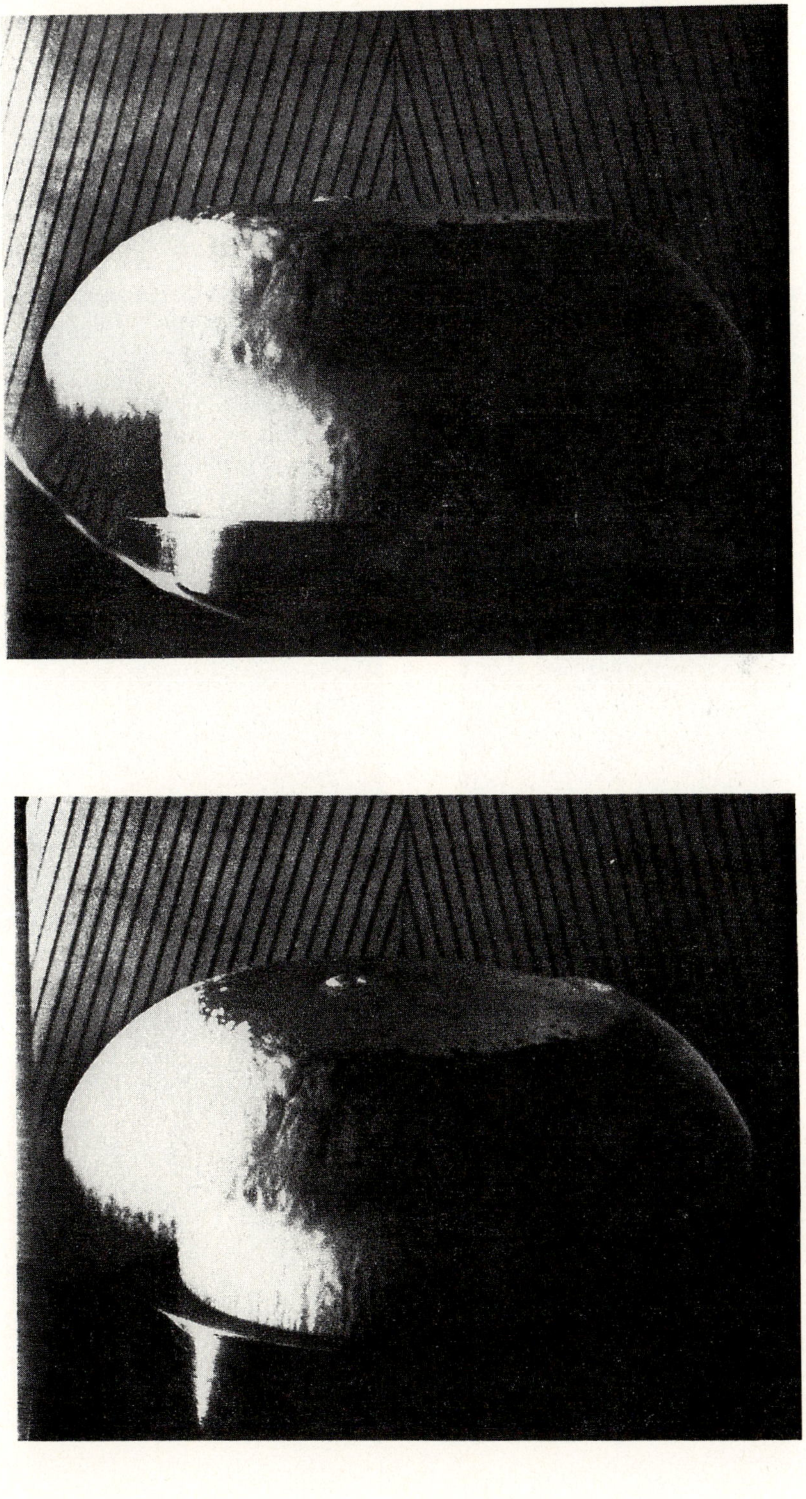

(a)

(b)

Fig. 5

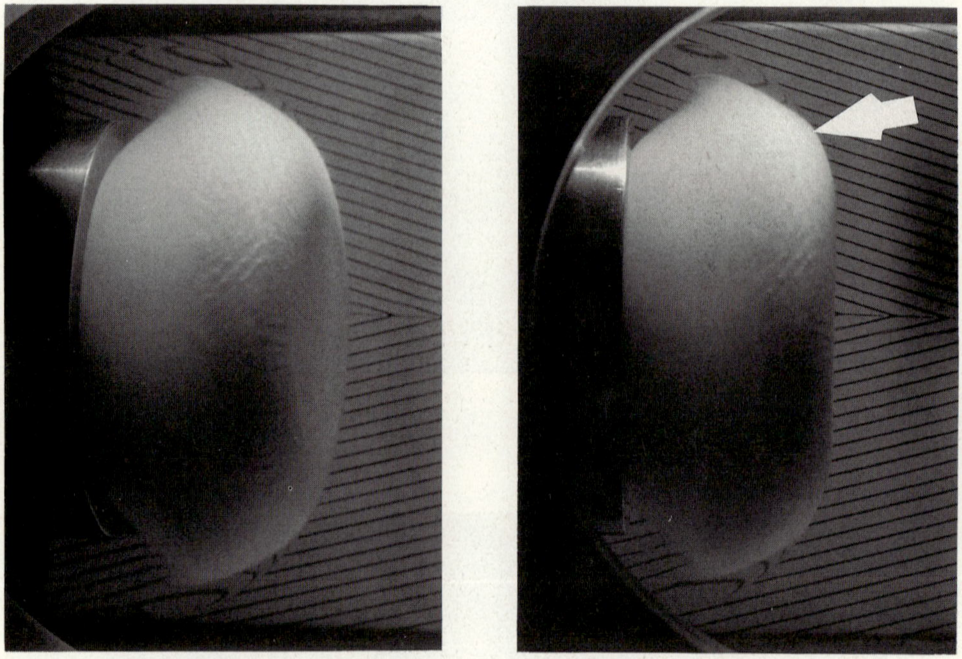

(a) (b) Fig. 6

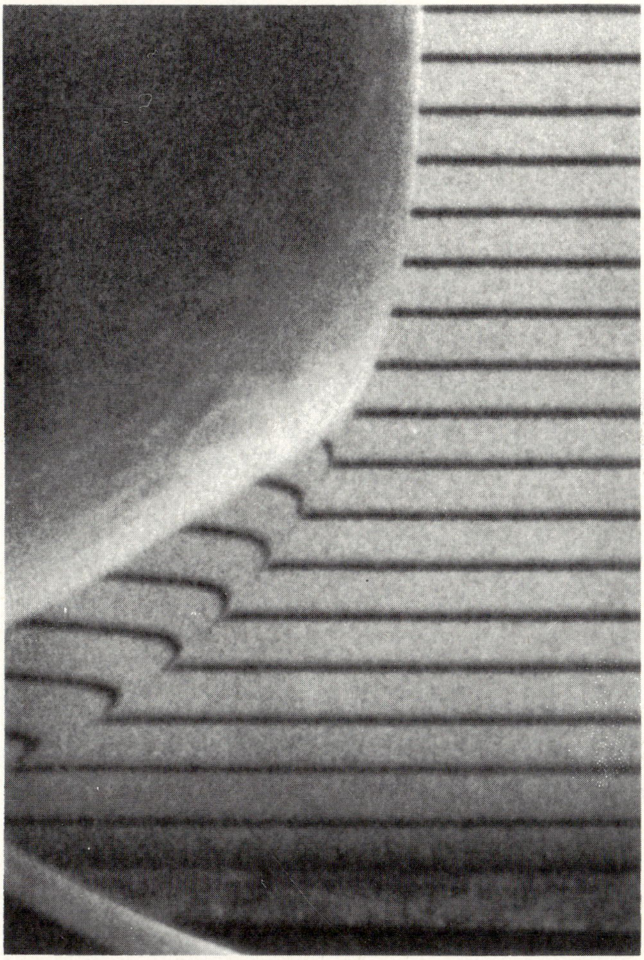

Fig. 7

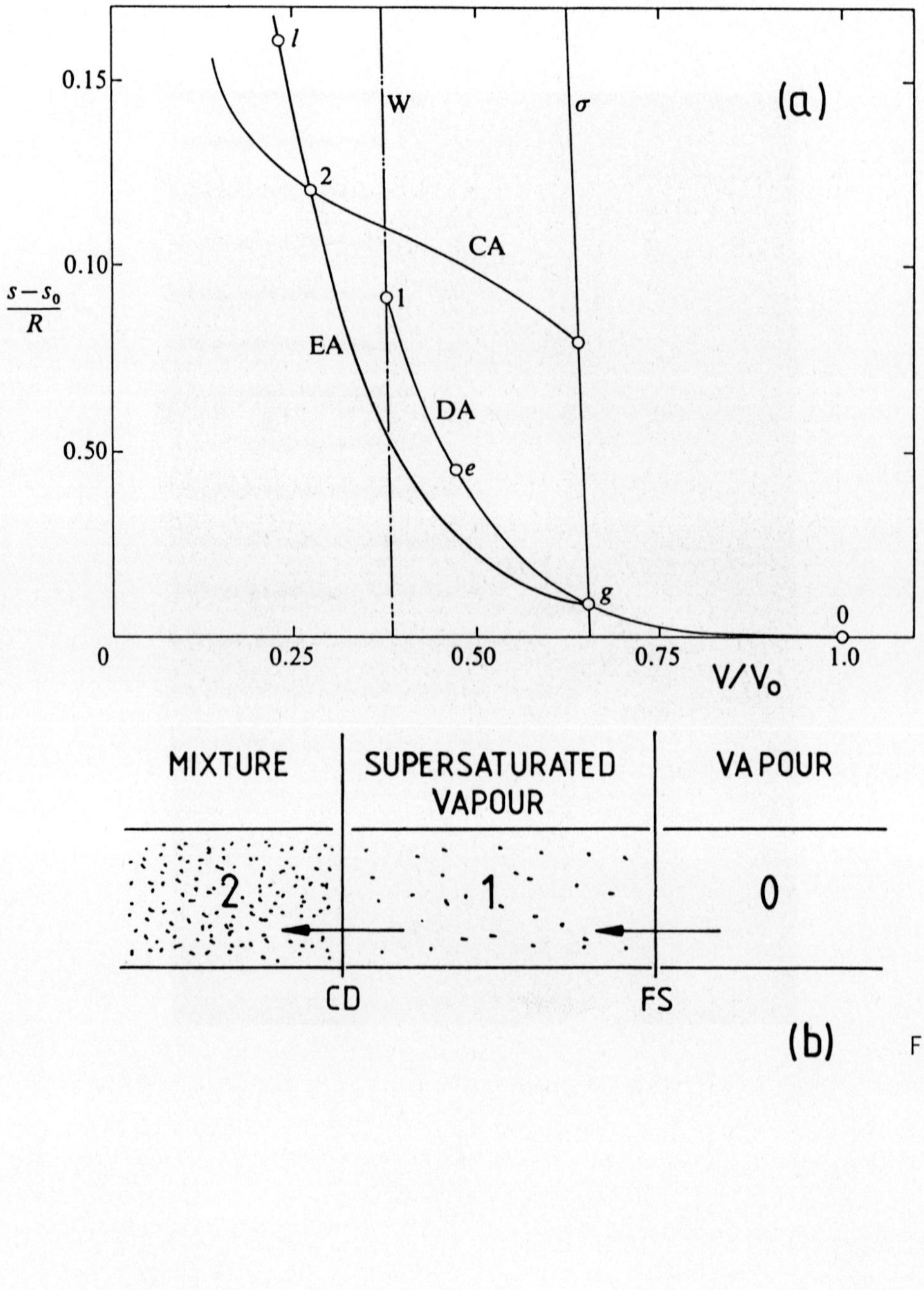
0.15
0.10
0.50
$\frac{s-s_0}{R}$
0
0.25
0.50
0.75
1.0
V/V₀
(a)
W
σ
l
2
CA
1
EA
DA
e
g
0
MIXTURE
SUPERSATURATED VAPOUR
VAPOUR
2
1
0
CD
FS
(b)

Fig. 8

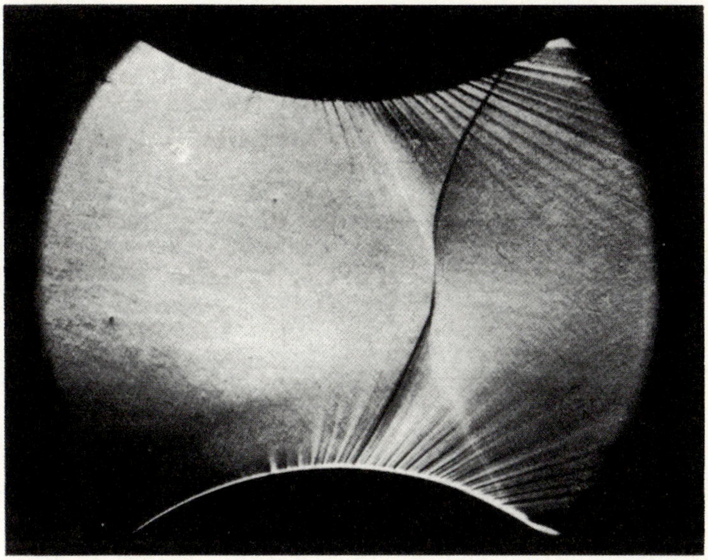

Fig. 9

Fig. 10

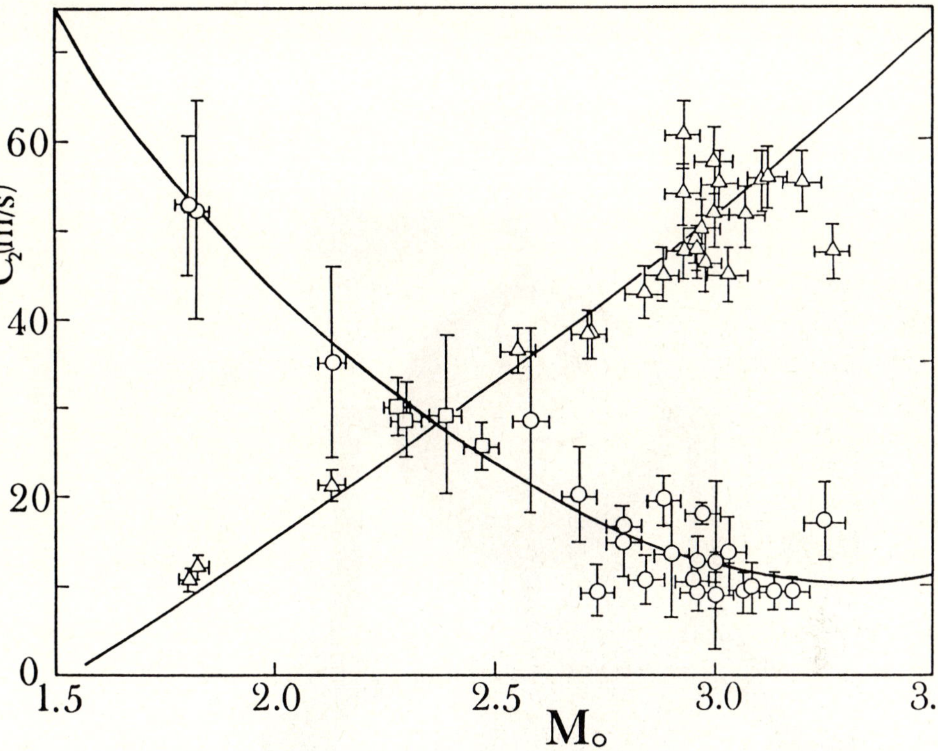

Fig. 11

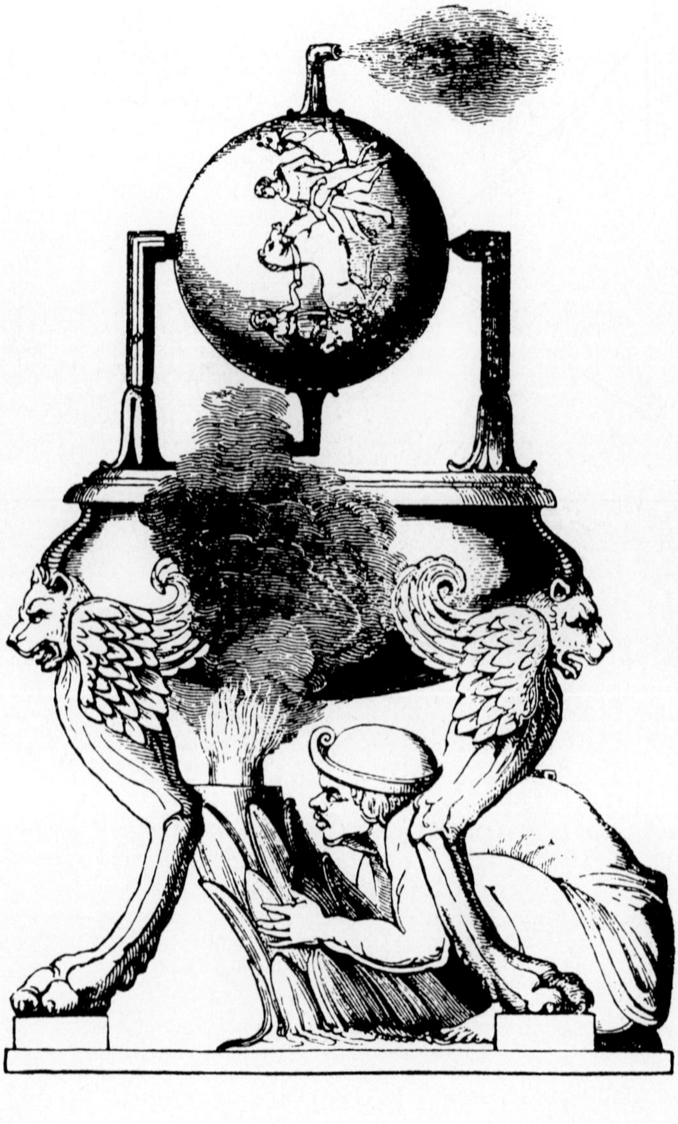

Fig. 12

Part II

MATHEMATICAL ANALYSIS

THE CONTINUOUS STRUCTURE OF DISCONTINUITIES

J. Glimm [1, 2, 3, 4]

Courant Institute of Mathematical Sciences
New York University
New York, N. Y. 10012

ABSTRACT

The regularization of discontinuities is discussed on the basis of molecular, computational and continuum considerations. Different regularization procedures may be implied by these distinct points of view. The mathematical motivation of regularization as an intermediate step in an existence proof is also of interest, but is not discussed in this paper.

1. Introduction.

Discontinuities in continuous media are defined by jump relations, which in turn are derived from the conservation laws of continuum physics. There are two types of physical mechanisms which provide a continuous structure internal to these discontinuities. The mechanisms are either molecular in origin or they have a basis in the continuum equations themselves.

This paper will develop these ideas and will illustrate both mechanisms by examples taken in part from the front tracking program pursued by the author and co-workers. We also discuss computationally motivated mechanisms for the regularization of discontinuities. It is a pleasure to acknowledge helpful conversations with P. Thompson.

1. Supported in part by the National Science Foundation, grant DMS - 8619856
2. Supported in part by the Applied Mathematical Sciences subprogram of the Office of Energy Research, U. S. Department of Energy, under contract DE-AC02-76ER03077
3. Supported in part by the Army Research Office, grant DAAG29-85-0188
4. Supported in part by the Air Force Office Office of Scientific Research AFSOR-88-0025.

2. Molecular Regularization Theories

An example of a direct molecular mechanism which generates the internal structure of a discontinuity is given by compressible reactive fluid dynamics. Part of the data for a compressible fluid problem is its equation of state, which specifies, for example, the internal energy e as a function of density ρ and temperature T, so that $e = e(\rho,T)$. There are thermodynamic constraints on e, which in turn influence the allowed discontinuous waves supported by the fluid equations. We focus here on phase transitions, considered as discontinuous fluid waves. A development of the relation between the equation of state and the allowed fluid waves for realistic materials subject to equilibrium thermodynamics is given in [12].

On the level of equilibrium thermodynamics, the width of a phase transition boundary (except near a critical point) is in the range of one or a few atoms or molecules. There is no physically meaningful internal structure to the phase boundary on the level of continuum mechanics and equilibrium thermodynamics. Common models for dynamic phase transitions use only equilibrium thermodynamic states. In the Stefan problem the interface is maintained at the phase transition temperature, while the front velocity is determined from a balance between the heat flux into the interface and the latent heat of the transition. The Mullins-Sekerka equations correct the melting temperature to account for front curvature, but are otherwise of the same nature. Both the Stefan and the Mullins-Sekerka equations give rise to sharp (unregularized) interfaces. We thus discuss nonequilibrium concepts and metastable states.

Extensive experimental studies of phase transitions starting from a metastable state have been performed [19]. The use of the ZND theory of detonations to describe phase transitions is presented in [13]. In order to incorporate nonequilibrium concepts into the description of the fluid flow, it is necessary to include an extra dependent variable, describing the relaxation process of the nonequilibrium ensemble towards equilibrium. This variable is then treated on the same level as a chemical reaction progress variable, and the equations have the same form as those of a chemically reacting fluid. Thus we are assuming that the departure from equilibrium occurs in a one dimensional subspace of allowed nonequilibrium states. This nonequilibrium state is then defined by a constraint on the thermodynamic ensemble. Subject to this constraint, all other thermodynamic variables are at equilibrium. Related to this discussion is the proposal [5] to use the Landau-Ginsberg equations as a model for dynamic phase transitions. The Landau-Ginsberg equations also contain an additional dependent variable: the order parameter.

The main molecular change associated with either chemical reactions or phase transitions is the breaking, forming or rearranging of chemical or molecular bonds. Combustion is exothermic, while dynamic phase transitions from a metastable state yield lower Gibbs free energy. In an unconstrained ensemble (with no constraint imposed by a reaction progress variable), each is a transition from a metastable state to a stable one. Perhaps the most significant difference between chemical reactions and phase transitions is that the chemical bonds are between charged particles (ions) and hence much stronger, so that the reaction energies are typically larger than in the case of phase transitions.

Nonequilibrium thermodynamics has also been investigated as a mechanism to produce internal structure in shock waves. The molecular phenomena here is the excitation of vibration modes in a diatomic molecule. A shock wave will excite these modes and induce a nonequilibrium partition of energy. The relaxation process then gives an internal structure and a finite width to the shock wave. A more common regularization of shock waves is via viscosity. Depending on the physical parameters, either or both of these two distinct mechanisms may be important. A study of relaxation effects in conservation laws is contained in [10].

Specifically, for the nucleation reaction to initiate conversion of a metastable phase ("A") to a stable phase ("B") we propose as a modeling approximation the reaction

$$A + B \to 2B \tag{2.1}$$

and the corresponding Arrhenius rate law

$$\dot{\lambda} = \lambda(\lambda_{eq} - \lambda)\, K \exp\left(\frac{-A}{T - T_c}\right) \equiv R_\phi\,, \tag{2.2}$$

with λ_{eq} the mass fraction in the phase B at equilibrium. For contrast, it is common to model combustion by the reaction

$$A \to B \tag{2.3}$$

and the rate law

$$\dot{\lambda} = (1 - \lambda)^\delta\, K \exp\left(\frac{-A}{T - T_c}\right) \equiv R_{comb}\,. \tag{2.4}$$

This rate law is modified by the condition $R = 0$ for $T < T_c$, to eliminate what is known as the cold boundary problem, or in other words, to make the unreacted state stable, rather than metastable. Here δ gives the order of the reaction, A is the activation energy, T_c is the ignition temperature and K is a multiplying factor in the rate law.

We note that Rabie Fowles and Fickett [13] use isothermal equations, so that the equation of state does not depend on temperature, and they use the simplified rate law

$$R = \lambda_{eq} - \lambda .$$

These authors comment that the explosive phase transitions they are discussing are driven by volume changes rather than by heat release. A more general concept of thermicity, which combines heat release and volume change, can be used to describe the driving mechanism of explosives. A more fundamental definition of the rate law, in terms of partial derivatives of the Gibbs energy, is also given in [13].

In either case, the Euler equations of a reactive fluid have the form

$$U_t + F(U)_r = C + G , \tag{2.5}$$

where $U = (\rho, m, E, \lambda)$ = (density, momentum, energy, reaction progress), F is the corresponding flux and C and G are chemical and geometrical source terms. The geometrical source terms arise from a cylindrical or spherical geometry, for example and the chemical source terms arise from the reaction rate equation (2.2) or (2.4). See [1,2,11,20].

There are two possible discontinuous transitions for the conservation laws (2.5), starting from an unreacted or metastable state $\lambda = 0$. The only fixed points of the rate equation are $\lambda = 0$ and $\lambda = \lambda_{eq}$, so the transitions are either of the form $\lambda = 0 \to \lambda = 0$ or $\lambda = 0 \to \lambda = \lambda_{eq}$. The allowed final states for the $\lambda = 0 \to \lambda = 0$ transition define the unreacted Hugoniot and they correspond to a shock wave through the metastable state defined by $\lambda = 0$. The allowed final states for the $\lambda = 0 \to \lambda = \lambda_{eq}$ transition correspond to transition to a thermodynamic equilibrium, i.e. a chemical reaction or a phase transition to a stable state. These states define the reacted Hugoniot. For the reactive Hugoniot, the jump relations involve the energy released by relaxation to a thermodynamic equilibrium state, as well as the volume change associated with the phase transition. According to the standard theory of combustion, the allowed states on the reactive Hugoniot fall into two branches: the detonation branch and the deflagration branch. Both corresponding types of phase transitions are known experimentally.

An important difference between combustion and phase transitions is the different nature of the physically selected detonation solutions [13]. Both combustion and metastable phase transitions depend essentially on two length scales [13]. As these length scales are set to zero, their ratio is an essential parameter. For this reason, the Riemann problem for the Euler equations does not have a unique solution. For phase transitions, the two length scales are the viscous length scale, which sets the width of the shock wave and a reaction rate, which sets the width for the deflagration, reaction, or phase transition wave.

The wave structure for the transition from a metastable state to an equilibrium state can have several distinct forms, depending on the state behind the transition. For illustration, we consider an unsupported wave, which is not driven by downstream pressure. Then the wave structure [13] consists of a weak detonation wave in which some reaction takes place. This wave connects the metastable pure phase state to a partially reacted mixed phase equilibrium state which is a critical point for the ODE's defining the traveling wave structure. A second wave, which is essentially a moving phase transition (deflagration), moving at a slower speed connects this critical point to the final equilibrium mixed phase or pure phase state. In the unsupported case, it ends at a CJ (sonic) point and a rarefaction wave can be attached to it.

The weak detonation wave is supersonic from behind, and thus does not satisfy the Lax admissibility condition. On this basis, the wave structure appears to be distinct from that analysed in the regularized van der Waals theory [16,17] of phase transitions. In the latter theory, a one parameter family of solutions to the Riemann problem is found, formed from right and left moving rarefactions or Lax shocks, and from moving phase boundaries. Furthermore the deflagration (phase transition) wave of [13] is supersonic relative to the sound speed of the mixed region it is located in, and ends at a sonic point in the unsupported case.

There is a considerable similarity between the qualitative aspects of detonation waves and of explosive phase transitions. There are also some interesting differences [13]. Regular crinkled 2-dimensional criss-crossing patterns in the wave front occur in both cases, and have been associated with surface instabilities and formation of Mach stems for the detonation case. Heterogeneous initiation sites (hot spots or nucleation sites) are an important feature of the reaction zone structure in both cases.

Front tracking has been applied to the computation of detonation problems [3]. It avoids (through greater dependence on mathematical analysis) the problem of grid resolution within the combustion zone, a problem which will be discussed below. Additional modeling information was required for this method, specifically the dependence of the detonation speed on the wave front curvature. This information has now been supplied [4,9,18]. Letting δD be the curvature correction to the wave speed, κ the curvature and δ the order of the reaction in (2.4), we have

$$\delta D = c_1\kappa \,, \qquad \delta < 1 \,,$$

$$\delta D = c_2\kappa \ln(\kappa) + c_3\kappa \,, \qquad \delta = 1 \,,$$

$$\delta D = c_4\kappa^{\frac{1}{\delta}} \,. \qquad 1 < \delta \,.$$

The reactive fluid equations, as given above, predict final reaction states, wave speeds and reaction zone widths for detonations, as has been verified by extensive computational and experimental studies [2,11]. Comparison between theory and experiment also requires a quantitatively correct equation of state [11], for which even the physically meaningful portion (the stable and the metastable points up to the Wilson line of instantaneous nucleation of the phase transition) of the van der Waals equation of state is completely inadequate in the case of detonations.

We observe more generally that for a regularization of a discontinuous wave to be correct according to the standards of science, it must not only produce a smooth traveling wave; it must also duplicate or model the mechanisms of physics and it must agree with experimental data. According to these standards, we propose the reactive fluid equations as offering a regularization of at least the rapid (explosive) branch of the dynamic phase transitions, following the theoretical work of [13] from 1979, as sketched above.

Both [13] and the regularized van der Waals theory of phase transitions build admissibility around traveling wave solutions. Essential differences between these two theories are that the details of the regularized equations are different and in the case of [13] have a satisfactory basis in physics; moreover the resulting wave structures for solutions of the Euler equations have qualitatively different features. Thompson et al [19] note differences between the wave structures known to experimentalists and those predicted by the regularized van der Waals theory.

Perhaps the simplest way to see the difference between the regularized van der Waals theory and the conclusions of theoretical and experimental physics is as follows: In the regularized van der Waals theory, there is no concept of an equilibrium mixed phase state (e.g. 50% ice, 50% water), because the Maxwell line of equilibrium states has been removed from the theory. Waves may start and end only at a stable or metastable single phase state. It follows that certain waves reported by theoretical and experimental physics are missing in the regularized van der Waals theory. These missing waves may arise in Riemann problems even if the left and right states are (stable or metastable) single phase states. The van der Waals theory of necessity avoids these missing waves, and then picks a qualitatively different one of the nonunique Riemann solutions from that reported in the physics literature.

3. Computationally Motivated Regularization Theories

The mathematical regularization of a discontinuity, independent of its physical basis, is also of interest. Numerical computations encounter many significant discontinuities. In the method of front tracking, these discontinuities are explicitly and correctly handled, and weak derivatives (jump conditions) are explicitly imposed. For finite differences, however, this is not the case, and solution values "within" the discontinuity are employed. The effect of a numerical discretization error for the finite difference method is typically to add to the equation a new term which is a polynomial $P(\frac{\partial}{\partial x}, \cdots)U$ of higher order in the derivatives. This polynomial then (may) define a numerical regularization, and an investigation of the mathematical properties of the regularization may shed light on either the correctness or on possible problems associated with the numerical procedures. As an example, it has been shown [6] that excess diffusion (under-resolution of the internal ZND wave structure) may cause significant errors in the detonation wave speed. Moreover, in the case of deflagrations, the flame speed is known to depend on the Prandtl number, which is the ratio of heat conduction to viscosity. It follows that these transport coefficients, or at least their ratio, must be correctly represented and not dominated by numerical artifacts, in the internal resolution of a numerically determined deflagration.

The viscosity capillarity regularization of the van der Waals fluid can be understood from the point of view of this section. The regularization has the form of an additional term $P(\frac{\partial}{\partial x}, \cdots)U$, added to the equations. It would be of interest to determine whether a deeper study of numerical phase transitions is called for. For example are numerically under-resolved phase transitions sensitive to artifacts of numerical algorithms? Moreover, the use of front tracking or similar interface methods for the numerical solution of moving boundary problems associated with phase transitions may be desirable.

4. Continuum Based Regularization Theories

A central challenge to continuum science is to develop a theory of continuum level microstructures and to integrate this theory into macroscopic continuum models. Turbulence is a special case of this problem, in which the microstructures are vortex filaments. In metallurgy, the microstructures are the defects, voids, impurities and grain boundaries which distinguish a typically real material from a perfect crystal. In favorable cases, macroscopic behavior can be determined numerically, for given response functions. In favorable cases, the response functions for pure materials (perfect crystals) can be determined from quantum mechanics. However there seem to be very few cases for which the continuum microstructures can first be solved and then their influence on the macroscopic materials (such as the constitutive relations for real materials) determined.

The continuum based regularization of discontinuities is a special case of this broader problem. In the case of phase transitions, the microstructures are dendrites, caused by a surface instability of a dynamic phase boundary, growing into a region occupied by metastable material. On suitable length scales, the variable phase in the dendrite region can be averaged, to produce a smooth transition zone, for example of "mush" between ice and water. This procedure leads to two phase flow equations, with a variable $\lambda(x,t)$ to represent the local average fraction of ice in a small volume about the space time point x, t.

It would be desirable, but it is not presently feasible, to proceed from a microscopically correct description of phase transitions (based on sharp interfaces or on relaxation rates for metastable states as discussed above) to the dendrites, then to statistical theories of ensembles of dendrites, and finally to continuum two phase flow equations.

Here we present some details, as well as some recent progress, in a related interface regularization program. The problem is that of mixing or entrainment across a fluid interface, when the molecular diffusion process is negligible, and continuum (but still microscopic, i.e. small with respect to an interfacial mixing layer) events dominate. More specifically, we are concerned with acceleration driven or Rayleigh-Taylor unstable interfaces. In view of the relation of this problem to the broader problem of microstructures for continuum mechanics, even intermediate progress is of considerable interest.

We mainly emphasize our general conclusions, which may be relevant to other continuum based regularization problems. Such similar theories include vortex roll up, Saffman-Taylor fingering in Hele-Shaw cells and the Richtmyer-Meshkov instability of impulsively accelerated interfaces.

A common feature of the interface substructure is the formation of distinctive modes, which grow to large amplitude and eventually interact with other modes. A typical interaction leads to mode merger, and thus to larger length scales; mode splitting can also occur but seems to be less important. We are interested, among other topics, in the outer envelope of the mixing zone (i.e. the width of the regularized interface). Because of the transfer of information to larger length scales, there is a dimensional renormalization in the width of the mixing zone. In the special case of a Rayleigh-Taylor unstable mixing zone, the width w has the form

$$w = w(t) = \alpha g t^2$$

as a function of the time t and the gravity or acceleration g. α is a constant, and it is one of the goals of the theory to determine α.

Our first main accomplishment was to obtain a high quality numerical solution of the single mode problem (one bubble rising under gravity) in a channel, in the compressible case [7]. Moreover, we were in agreement with a large body of previous work for the incompressible limit for this problem.

By analyzing experiments of Read [14], it was determined [8] that the single bubble velocity in a channel did not agree with bubble velocities in complex flow patterns, and the disagreement was by a factor of 2 to 3. This result should be a warning to workers on related problems, as there is a natural tendency to expend considerable attention on the single mode problem. We now believe that the single mode problem has to be solved in a self-consistent way, in a background field defined by the other modes. More precisely, the velocity of a single bubble depends on its position relative to its neighbor bubbles. It seems that correlations develop among neighboring bubble positions so that after elimination of initial transients, the influence of neighboring bubbles can be predicted, and terminal bubble velocities obtained. These correlations may depend on surface tension, or the tendency to have bubble splitting, and do not seem to be universal in their quantitative aspects.

A statistical model of Sharp and Wheeler [15] was used to study ensembles of many bubbles, with simple input from the one body (single mode) and two body (bubble merger) problems. We took this information from the experiment [14], and as a result obtained [8] reasonable agreement between the experimental uniform acceleration rate, α_{exp}, and the model acceleration rate, α_{SW},

$$\alpha_{\text{exp}} \approx \alpha_{\text{SW}} .$$

Universal behavior of α_{exp} has been observed experimentally [14], and a related but distinct universal behavior for α_{SW} was discovered in the Sharp-Wheeler model [8].

A better understanding of the one and two body problems based on solutions of the compressible two fluid Euler equations, as well as a direct simulation of the multibubble problem, to give a direct computational determination of α, remain as important problems for future work.

References

1. Courant and Friedrichs, *Supersonic Flow and Shock Waves,* Springer-Verlag, New York, 1976.
2. W. Fickett and W. C. Davis, *Detonation,* Univ. of California Press, Berkeley, 1979.
3. B. Bukiet, "Applications of Front Tracking to Two Dimensional Curved Detonation Fronts," *SIAM J. Sci. Stat. Comp.*, vol. 9, pp. 80-99, 1988.
4. B. Bukiet, "The Effect of Curvature on Detonation Speed," *SIAM J. Appl. Math.*, To appear.
5. G. Caginalp, "An Analysis of a Phase Field Model of a Free Boundary," *Archive for Rational Mechanics and Analysis*, vol. 92, pp. 205-245, 1986.

6. P. Colella, A. Majda, and V. Roytburd, "Theoretical and Numerical Structure for Reacting Shock Waves," *Siam J. Sci Stat Comp*, vol. 7, pp. 1059-1080, 1986.

7. C. Gardner, J. Glimm, O. McBryan, R. Menikoff, D. H. Sharp, and Q. Zhang, "The Dynamics of Bubble Growth for Rayleigh-Taylor Unstable Interfaces," *Phys. of Fluids*, In Press.

8. J. Glimm and X.L. Li, "On the Validation of the Sharp-Wheeler Bubble Merger Model from Experimental and Computational Data," *Phys. of Fluids*, To appear.

9. J. Jones, "The Spherical Detonation," *Comm. Pure Appl. Math.*, To appear.

10. T.-P. Liu, "Hyperbolic Conservation Laws with Relaxation," *Comm. Math. Phys.*, vol. 108, pp. 153-175, 1987.

11. C. Mader, *Numerical Modeling of Detonations*, Univ. of California Press, Berkeley, 1979.

12. R. Menikoff and B. Plohr, "Riemann Problem for Fluid Flow of Real Materials," Los Alamos preprint LA-UR-2259, 1987.

13. R. L. Rabie, G. R. Fowles, and W. Fickett, "The Polymorphic Detonation," *Physics of Fluids*, vol. 22, pp. 422-435, 1979.

14. K. I. Read, "Experimental Investigation of Turbulent Mixing by Rayleigh-Taylor Instability," *Physica 12D*, pp. 45-48, 1984.

15. D. H. Sharp, "Overview of Rayleigh-Taylor Instability," *Physica 12D*, pp. 3-17, 1984.

16. M. Slemrod, "Admissibility Criteria for Propagating Phase Boundaries in a van der Waals Fluid," *Arch. Rat. Mech. Anal.*, vol. 81, pp. 303-319, 1983.

17. M. Slemrod, "Dynamic Phase Transitions in a Van Der Waals Fluid," *J Diff. Eq.*, vol. 52, pp. 1-23, 1984.

18. S. Stewart and J. B. Bdzil, "The Shock Dynamics of Stable Multidimensional Detonation," *J. Fluid Mech.*, To appear.

19. P. Thompson, G. Carofano, and Y.-G. Kim, "Shock Waves and Phase Changes in a Large Heat Capacity Fluid Emerging from a Tube," *J. Fluid Mech.*, vol. 166, pp. 57-92, 1986.

20. Forman Williams, *Combustion Theory*, Addison-Wesley Co., Reading, 1965.

RIEMANN PROBLEMS INVOLVING UNDERCOMPRESSIVE SHOCKS

Michael Shearer and Stephen Schecter
Department of Mathematics
North Carolina State University
Raleigh, North Carolina 27695.

1. Introduction. Recent work on nonstrictly hyperbolic conservation laws has highlighted the importance of undercompressive shocks [7,9]. These are shock wave solutions that possess viscous travelling wave profiles and yet according to the classical theory of strictly hyperbolic systems have too few characteristics entering the shock. In [7], the Riemann problem was solved for a special system of nonstrictly hyperbolic conservation laws:

$$\begin{aligned} u_t + (-u^2 + v^2)_x &= 0 \\ v_t + (2uv)_x &= 0. \end{aligned} \tag{1.1}$$

Solution of the Riemann problem for (1.1) was facilitated by the S_3 symmetry in (1.1) between u, v and $1-u-v$. In particular, undercompressive shocks for (1.1) were characterized as involving jumps in (u, v) only on the lines of symmetry for (1.1).

In [3] it was shown that 2×2 systems of nonstrictly hyperbolic conservation laws with an isolated umbilic point could be reduced to the normal form

$$\begin{aligned} u_t + (au^2 + 2buv + v^2)_x &= 0 \\ v_t + (bu^2 + 2uv)_x &= 0, \end{aligned} \tag{1.2}$$

with $a \neq 1 + b^2$. Note that (1.2) reduces to (1.1) when $a = -1,\ b = 0$.

For $a > 3b^2/4$, the Riemann problem was solved in [4], and does not involve undercompressive shocks. The main stumbling block to understanding the remaining range of (a, b) has been the lack of a workable characterization of undercompressive shocks. We use a result of Chicone [1] to describe admissible shocks, by which we mean shocks with viscous profiles for the parabolic system

$$\begin{aligned} u_t + (au^2 + 2buv + v^2)_x &= \epsilon u_{xx} \\ v_t + (bu^2 + 2uv)_x &= \epsilon v_{xx}, \end{aligned} \tag{1.3}$$

with $\epsilon > 0$. In §3, we summarize the key observations that lead to a solution of the Riemann problem for $a < 3b^2/4$.

In §4, we consider perturbations of the nonlinearity in (1.2) by cubic and higher order terms. We show how Melnikov's method from dynamical systems is used to characterize undercompressive shocks. At the heart of the analysis is a new study of the simultaneous bifurcation of equilibria and heteroclinic orbits for a vector field in the plane.

In §5, we indicate preliminary results on solving the Riemann problem for these equations with cubic nonlinearities. The results are stable to higher order perturbation.

2. Preliminaries.

The function

$$U(x,t) = \begin{cases} U_- & \text{if } x < \text{st} \\ U_+ & \text{if } x > \text{st} \end{cases} \tag{2.1}$$

is a shock wave solution of the 2×2 system of conservation laws

$$U_t + F(U)_x = 0 \quad U \in \mathbb{R}^2) \tag{2.2}$$

if $U_\pm$ and the shock speed s satisfy the Rankine-Hugoniot conditions:

$$F(U_+) - F(U_-) - s(U_+ - U_-) = 0 \tag{2.3}$$

Characteristic speeds for (2.1) are eigenvalues $\lambda_1(U) < \lambda_2(U)$ of the Jacobian matrix $dF(U)$. (We shall only consider hyperbolic equations, for which the characteristic speeds are real.) For a piecewise constant shock (2.1), the characteristics are straight on either side of the shock, and have speeds $\lambda_k(U_\pm)$. The kth family of characteristics enters the shock (2.2) on the left if its speed $\lambda_k(U_-)$ on the left is greater than the shock speed s : $\lambda_k(U_-) > s$. Similarly, the kth family of characteristics enters the shock on the right if $s < \lambda_k(U_+)$. A shock satisfies the Lax entropy condition if one family of characteristics enters the shock on both sides of the shock. We shall speak of such shocks as being Lax shocks or compressive shocks. There are slow and fast Lax shocks. Slow shocks have both slow characteristics entering the shock, while fast shocks have both fast characteristics entering the shock. Shocks that are both fast and slow in this sense are called overcompressive shocks [4]. (For (1.2), overcompressive shocks occur only for a $> 3b^2/4$.) In this paper, we also consider shocks that are not compressive because both families of characteristics pass through the shock. That is, each family of characteristics enters the shock on only one side. Not all of these noncompressive shocks are physically reasonable, and indeed initial value problems for (2.1) do not in general have unique solutions in the class of weak solutions with compressive and noncompressive shocks. A suitable criterion for selecting those noncompressive shocks that are physical is to insist that the shock can be approximated by a smooth solution of a regularized version of (2.2) obtained by including dissipative effects in the derivation of the equations of motion. Here, we consider the following representative form of such dissipative terms. A shock (2.1) is called admissible if there is a traveling wave solution $U = U((x - st)/\epsilon)$ of the system

$$U_t + F(U)_x = \epsilon U_{xx} \tag{2.4}$$

satisfying $U(\pm\infty) = U_\pm$ and $U'(\pm\infty) = 0$. Carrying out the differentiations in (2.4) and integrating once leads to a system of ordinary differential equations, with parameters s and U_-:

$$U' = F(U) - F(U_-) - s(U - U_-) \tag{2.5}$$

Note that U_- is an equilibrium for (2.5) for all s. A point U_+ is a second equilibrium for (2.5) if and only if $(U_\pm, s)$ satisfies the Rankine Hugoniot conditions (2.3). For a slow shock, U_- is an unstable node, and U_+ is a saddle point, whereas for a fast shock, U_- is a saddle point and U_+ is a stable node. If (2.1) is a compressive fast or slow shock, and U_-, U_+ are sufficiently close together, then there is a trajectory for (2.5) joining U_- to U_+. Thus, weak Lax shocks are admissible. However, if U_-, U_+ are not close, then there may be no trajectory for (2.1) joining the two equilibria. In this case, we regard the Lax shock as being inadmissible. If the shock is noncompressive, with both families of characteristics passing through the shock, then both U_- and U_+ are saddle points for (2.5), and the shock is undercompressive if there is a trajectory between these two saddle points. The trajectories between equilibria are referred to as heteroclinic. We shall sometimes refer to (2.1) as a shock $U_- \to U_+$ with speed s. The Hugoniot locus $H(U_-)$ consists of all points U_+ such that $U_- \to U_+$ is a shock with some speed s.

The Riemann problem for (2.1) is the initial value problem with initial data

$$U(x,0) = \begin{cases} U_L & \text{if } x < 0 \\ U_R & \text{if } x > 0 \end{cases} \tag{2.6}$$

The Riemann problem is solved by looking for solutions $U = U(x/t)$ that are constant except across shocks and through rarefaction waves. Centered rarefaction wave solutions of (2.1) are continuous piecewise smooth solutions $U(x/t)$. Substitution into (2.1) leads to the equations $x/t = \lambda_k(U)$ and $U' = r_k(U)$, where $r_k(U)$ is the right eigenvector of $dF(U)$ corresponding to the eigenvalue $\lambda_k(U)$, normalized so that $\nabla\lambda_k(U)\cdot r_k(U) = 1$. Thus, $U(\xi)$ lies on an integral curve of $r_k(U)$. The normalization of $r_k(U)$ makes sense when system (2.1) is genuinely nonlinear. Genuine nonlinearity fails when $\nabla\lambda_k(U)$ is orthogonal to the corresponding eigenvector of $dF(U)$. Such points generically form curves called inflection loci in the U-plane. (In some systems, genuine nonlinearity fails everwhere, but such linear degeneracy will not be discussed here.) For system (1.2), the inflection loci are straight lines through the origin. For $a < 3b^2/4$, there are three inflection loci, including multiplicity, whereas for $a > 3b^2/4$, there is just one inflection locus $bu + v = 0$. (See [3] for details.) Since the characteristic speed within a rarefaction wave must increase through the wave with increasing x, $\lambda_k(U)$ is monotonic increasing along the corresponding portion of the integral curve of $r_k(U)$. With this orientation, the integral curves are called rarefaction curves. The orientation reverses across inflection loci. Thus, inflection loci limit the strength of rarefaction waves.

In [7], undercompressive shocks are discussed in detail for system (1.1). We summarize the results here. First, let $v_- = 0$. Then $(u_+, 0)$ satisfies the Rankine-Hugoniot conditions (2.3) for any u_+, with $s = -(u_+ + u_-)$. Let $u_- < 0$. Then the shock wave

$$(u,v) = \begin{cases} (u_-, 0) & \text{if } x < \text{st} \\ (u_+, 0) & \text{if } x > \text{st} \end{cases} \tag{2.7}$$

is a slow shock for $u_+ > -3u_-$, an undercompressive shock if $-u_-/3 < u_+ < -3u_-$, and a fast shock if $u_- < u_+ < -u_-/3$. Each of these shocks is admissible. Now keep s fixed in the range $2u_- < s < -2u_-/3$ for undercompressive shocks, and set $u_+ = u_- - s$. For $v_- \neq 0$, near zero, there are no undercompressive shocks with left state (u_-, v_-) and speed s. That is, although there is a saddle point near $(u_+, 0)$, and (u_-, v_-) is a saddle point, there is no trajectory joining the two saddle points. Undercompressive shocks are indicated in Fig. 1 by Σ.

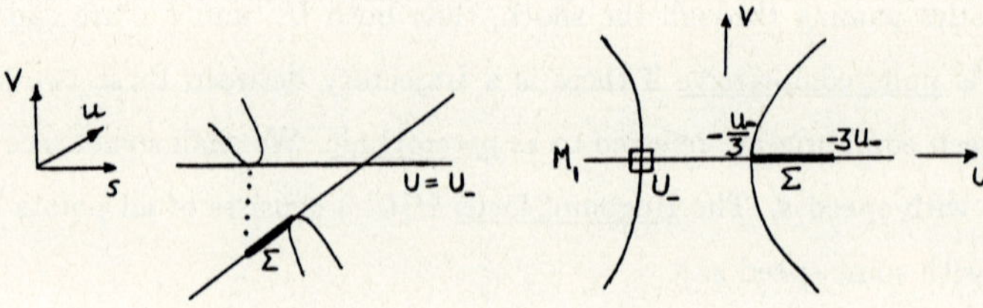

Bifurcation Diagram Hugoniot Locus

Fig. 1. Shocks for (1.1); $U_- = (u_-, 0)$, $u_- < 0$.

3. Quadratic Nonlinearities. In this section, we describe the key ideas in solving the Riemann problem for system (1.2) with $a < 3b^2/4$. System (2.5) becomes

$$\begin{aligned} u' &= au^2 + 2buv + v^2 - au_-^2 - 2bu_-v_- - v_-^2 - s(u - u_-) \\ v' &= bu^2 + 2uv - bu_-^2 - 2u_-v_- - s(v - v_-) \end{aligned} \tag{3.1}$$

Let $f : \mathbb{R}^2 \to \mathbb{R}^2$ be a cubic polynomial and let $Q = \nabla f$. Then Q is called a quadratic gradient vector field. Note that the right hand side of (3.1) is a quadratic gradient vector field.

THEOREM 1. *(Chicone, [1]) Let $Q : \mathbb{R}^2 \to \mathbb{R}^2$ be a quadratic gradient vector field. Then every trajectory joining two saddle point equilibria of the system $U' = Q(U)$ lies on an invariant straight line.*

Invariant straight lines for (3.1) are referred to as medians. Each invariant line passes through the origin. Suppose (u_-, v_-) lies on an invariant line. Then if $v_- = mu_-$, the invariant line is $v = mu$, where m satisfies

$$m^3 + 2bm^2 + (a-2)m - b = 0. \tag{3.2}$$

Equation (3.2) gives the same invariant line for each s. It is easy to check that each invariant line has two equilibria for (3.1). For $a < 1 + b^2$, there are three invariant lines, but only for a $< 3b^2/4$ do the invariant lines contain two saddle point equilibria for an interval of values of s. For each (u_-, v_-) not on a median, there may be two equilibria for (3.1) on a median, for some value of s. These two equilibria are joined by a trajectory along the median, whether or not they are saddles. This observation is crucial to understanding which shock waves have viscous profiles. Let M_k, $k = 1,2,3$ denote the medians. Let U_L be a point not on any median. If U_- lies in the Hugoniot locus $H(U_L)$ of U_L, with shock speed s and U_- also lies on the median M_k, then there exists a second point U_+ in $H(U_L) \cap M_k$. If $U_L \to U_-$ is a slow shock with speed s, and U_- lies in the same component of $H(U_L)$ as U_L, then $U_- \to U_+$ is an undercompressive shock with the same speed s. If $U_- \in M_k$ lies on the slow rarefaction curve through U_L, then there is $U_+ \in M_k$ such that $U_- \to U_+$ is an undercompressive shock whose speed is $\lambda_1(U_-)$. Of course this shock has the slow family of characteristics parallel to the shock on the left, so it is on the boundary between undercompressive and compressive shocks. Any shock with U_- on the left and slower speed would not be possible in this construction.

To solve the Riemann problem, we adopt the strategy of [4,7]. For simplicity, we only work near the median M_1 that reduces to the u-axis when $b = 0$. We first construct the bifurcation diagrams representing solutions of the Rankine-Hugoniot conditions (2.3). As usual, s is the bifurcation parameter and u_-, v_- are treated as auxiliary parameters. On these diagrams, we indicate where there are admissible shocks. In working out this structure, we have the local saddle-to-node connections of fast and slow weak shocks, and the benefit of Chicone's result. The latter not only characterizes undercompressive shocks, it also tells us where the saddle-to-node connections break down. The bifurcation diagrams are given in Fig. 2, and their projections onto the (u, v) plane in Fig. 3. Consider U_- above M_1. In Fig. 2, the admissible portion S_1^* of the bifurcation diagram ends at a point Q', with $s < \lambda_-(U_-)$. The Lax entropy condition would include the portion between $Q'$and the plane $s = \lambda_-(U_-)$. Now consider the projection onto the (u, v)-plane in Fig. 3. Let P, Q be the points of intersection of $H(U_-)$ with M_1. Then Q is the projection of Q', and P is the projection of P', with the same value of s. In Fig. 4, we show the phase plane for values of U_+ near $U_+ = Q$ on the Hugoniot locus, to show how the saddle to saddle connection $P \to Q$ cuts off the node to saddle connection $U_- \to Q_-$. The pair U_-, Q_+ satisfies the Lax entropy condition, but the corresponding shock is not admissible because there is no heteroclinic orbit joining U_- to Q_+.

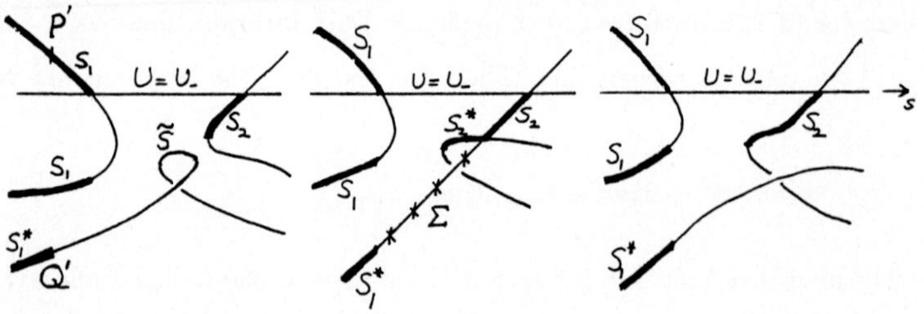

U_- above M_1 $\qquad$ U_- on M_1 $\qquad$ U_- below M_1

Fig. 2. Bifurcation diagrams for (1.2), $b \neq 0$.

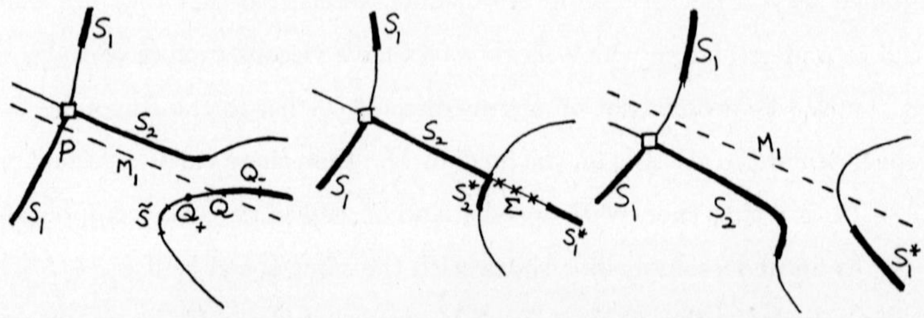

U_- above M_1 $\qquad$ U_- on M_1 $\qquad$ U_- below M_1

Fig. 3. Hugoniot loci for (1.2), $b \neq 0$

The portions of curves labelled $\tilde{S}$, S_2^* in Figs. 2,3 require some explanation. There is no connection from U_- to points on S_2^* because the unstable manifold from the saddle point U_- lies on the median. However, for each point U on S_2^*, there is a saddle point U_+ on Σ for the same value of s. U_- is joined to U_+ and U_+ is joined to U. Thus there is a shock $U_- \to U$ with speed s that is the limit of an undercompressive shock $U_- \to U_+$ and a fast shock $U_+ \to \tilde{U}$, as $\tilde{U}$ approaches U. This justifies the labeling of S_2^* as an admissible shock curve. The portion $\tilde{S}$ (which approaches S_2^* as U_- approaches M_1) is inadmissible (despite satisfying the Lax entropy condition). For points U_+ on $\tilde{S}$, the unstable manifold of U_- is unbounded.

In Fig. 5, we give the solution of the Riemann problem as U_L crosses the median M_1. From this, it is possible to construct the solution of the Riemann problem for all U_L, U_R. This will be done in a future paper [8]. The notation in Fig. 5 is the same as that in [4,7].

<u>4. Melnikov's method</u>. In this section, we summarize preliminary results concerning undercompressive shocks for perturbations of system (1.3) that include cubic flux functions. The procedure used is known as Melnikov's method. This provides a local characterization of saddle-to-saddle heteroclinic orbits that replaces Chicone's result, Theorem 1, which applies only to quadratic gradient flux functions.

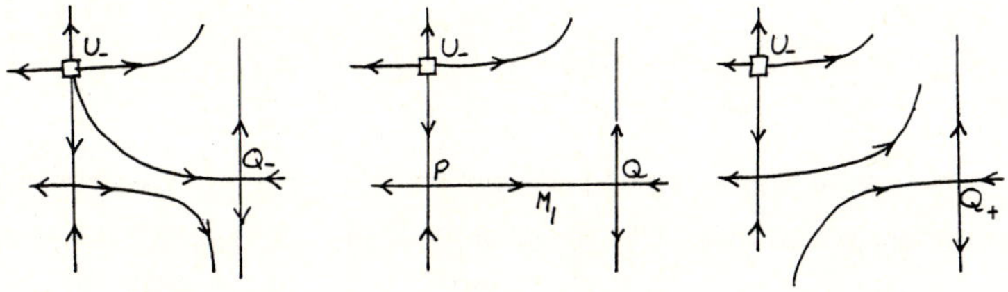

$U_+ = Q_-$ $U_+ = Q$ $U_+ = Q_+$

Fig. 4. Phase plane for (1.2), U_- above M_1.

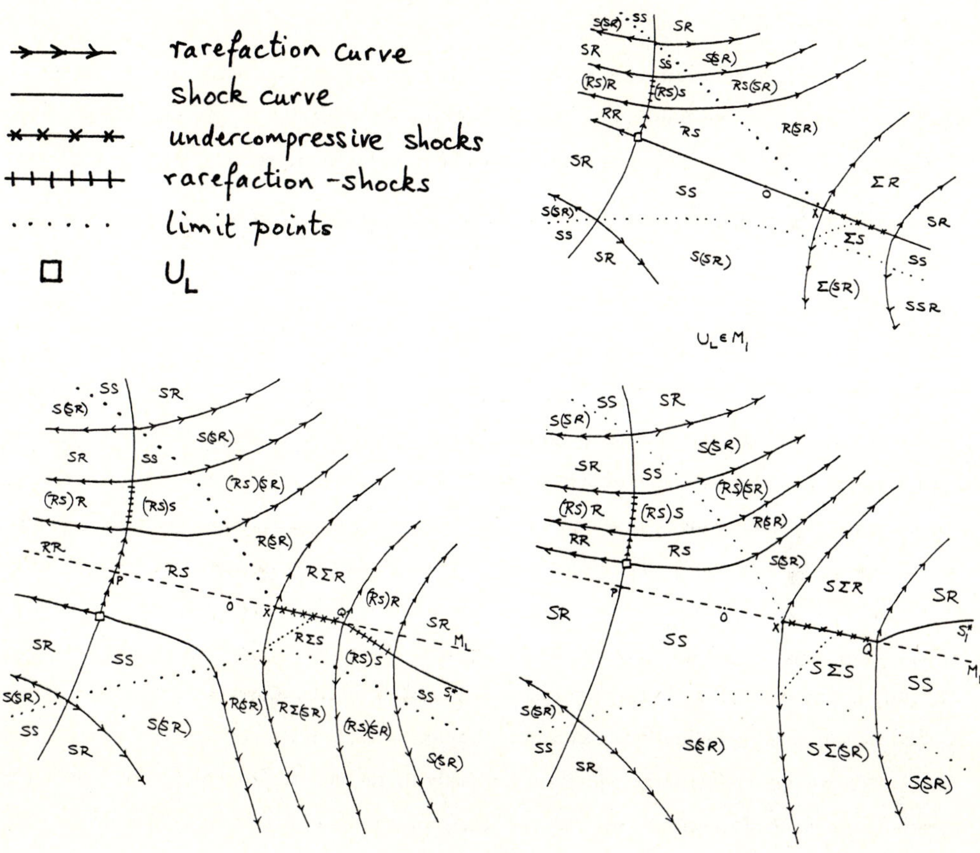

Fig. 5. Solution of Riemann problems for (1.2).

Consider the system

$$U_t + F(U)_x = \epsilon U_{xx} \tag{4.1}$$

with

$$F(u,v) = d(au^3/3 + bu^2v + uv^2) + C(u,v) \tag{4.2}$$

where

$$C(u,v) = \begin{pmatrix} A_{11}u^3 + A_{12}u^2v + A_{13}uv^2 + A_{14}v^3 \\ A_{21}u^3 + A_{22}u^2v + A_{23}uv^2 + A_{24}v^3 \end{pmatrix} \qquad (4.3)$$

Since we consider (4.2) to have a small cubic perturbation, the parameters A_{ij} are taken to be small. Equivalently, we might regard (4.2) as a truncated Taylor series expansion about an umbilic point, and restrict attention to a small neighborhood of the umbilic point. We find that near the undercompressive shocks of §2, the most significant parameter is A_{21}. To simplify formulae and some diagrams, we henceforth take $A_{ij} = 0$ unless $i = 2$, $j = 1$ and set $\epsilon = A_{21}$. We also set $u_- = -1/2$, without loss of generality.

The main result of this section is a description of admissible shocks with (u_-, v_-) near $(-1/2, 0)$ and (u_+, v_+) near one end point $(1/6, 0)$ of the section of the median representing undercompressive shocks for (1.1). We then use this information in §5 to solve the Riemann problem locally.

Set $a = -1$ in (4.2). We restrict attention to a neighborhood of the secondary bifurcation in Fig. 1. Then s is near $1/3$, and for small values of v_-, b, ϵ, the equilibrium (u_-, v_-) is a saddle point. When the parameters are all zero, including $\lambda = s - 1/3$, this saddle point is joined to a degenerate saddle point $(1/6, 0)$ by a heteroclinic orbit lying on the u-axis. The u-axis is the stable manifold for this equilibrium, while the center manifold is parallel to the v-axis. As the parameters vary, two things happen. The degenerate saddle point splits into up to three equilibria. The bifurcations take place parallel to the v-axis. The heteroclinic orbit also splits up. The unstable manifold Γ from (u_-, v_-) with $u > u_-$ may or may not be joined to another equilibrium. When Γ is joined, then we have an admissible shock. The other equilibria are generically saddles and nodes, with strongly stable invariant manifolds in the direction of the u-axis. These invariant manifolds are separated from Γ by a distance d depending on the parameters. To understand which shocks are admissible, we study the zero set of d. The first problem is that d should be a smooth function of parameters, so that we can look at the first few terms of its Taylor series, to characterize $d = 0$ as a manifold. To resolve this, we parameterize the equilibria smoothly by the coordinate v along the center manifold, together with some of the other parameters in the problem. Then a version of Melnikov's method [5] provides formulae for derivatives of d with respect to parameters. We summarize the results here. A complete analysis will be given elsewhere [6].

Treating $\lambda = s - 1/3$ as a distinguished parameter in the spirit of [2], we find the flow on the center manifold to have the normal form

$$\dot{x} = x^3 - \lambda x + \beta x^2 + \alpha \qquad (4.4)$$

Here, the state variable x and unfolding parameters α, β are given in terms of the original variables and parameters to first order by

$$x = v - \frac{1}{3}v_L + \frac{1}{9}b + \frac{1}{36}\epsilon$$
$$\alpha = \frac{4}{9}v_L - \frac{2}{27}b + \frac{7}{162}\epsilon$$
$$\beta = -3v_L + \frac{5}{6}b - \frac{31}{72}\epsilon .$$

The equilibria in (4.4) are parameterized smoothly by x, λ, β, with

$$\alpha = -\beta x^2 + \lambda x - x^3.$$

Then $d = d(x, \lambda, \beta, \epsilon)$ is parameterized smoothly. Since $d(0) = 0$, the first terms in the Taylor series of d are linear. The coefficients are calculated by Melnikov's method, which involves calculating improper integrals along the invariant manifold, with the result:

$$d = -\frac{4}{9}x + \frac{4}{243}\epsilon + \text{ higher order terms.} \tag{4.5}$$

Therefore, to first order, by the implicit function theorem, the set $d = 0$ is given by

$$x = \frac{1}{27}\epsilon + \text{ higher order terms.} \tag{4.6}$$

Next we turn to an interpretation of the result (4.6) that can be used to solve Riemann problems. For fixed (u_L, v_L) near the median M_1, with $u_L < 0$, let (u_-, v_-) vary along the slow wave curve $W_1(u_L, v_L)$. ($W_1(u_L, v_L)$ represents all shocks, rarefactions and rarefaction-shocks associated with the slow characteristic speed λ_1, and with (u_L, v_L) on the left of the wave in the (x,t)-plane. In this discussion, we only consider the portion of $W_1(u_L, v_L)$ that lies in a small neighborhood of (u_L, v_L).) This curve is transverse to M_1, so that u_- is roughly constant and v_- varies. For each v_- we have the Hugoniot locus $H(u_-, v_-)$ describing admissible and inadmissible shocks. To select the states that correspond to admissible shocks, we use bifurcation diagrams, labelling portions of the diagram as admissible by appealing to (4.6) and the theory of weak shocks.

In the discussion, we consider only portions of bifurcation diagrams near the secondary bifurcation of Fig. 1. For each fixed b and $\epsilon \neq 0$, we have a sequence of bifurcation diagrams varying with v_-. Moreover, the sequence of diagrams is not the same for each choice of b and ϵ. In fact, the (b, ϵ) plane divides into eight sectors (see Fig. 8), corresponding to four distinct classes of equations ($\epsilon > 0$ and $\epsilon < 0$ are similar). In the diagrams, various points are significant. Limit points are significant because the shock speed equals the characteristic speed, so the equilibrium(u_+, v_+) changes from a saddle to a node, generically. Also the point at which $d = 0$ is significant, not only because it describes a trajectory, but because it often represents where a saddle to node connection breaks down and is replaced by a saddle to saddle connection. Limit points coincide either at bifurcation points or at hysteresis points. These are drawn as curves $\mathcal{B}$ and $\mathcal{H}$ in Fig. 6. As we shall see, hysteresis points are insignificant for our purposes. In Fig. 6, $\epsilon < 0$ is fixed. We also show the locus $\mathcal{D}$ of points at which $d = 0$ coincides with a limit point somewhere in the bifurcation diagram, and the locus $\mathcal{D}'$ of values of α, β for which $d = 0$ occurs for the same value of λ as a limit point. As $\epsilon \to 0$, $\mathcal{H}$ and $\mathcal{B}$ remain fixed, but $\mathcal{D}$ and $\mathcal{D}'$ approach the β axis.

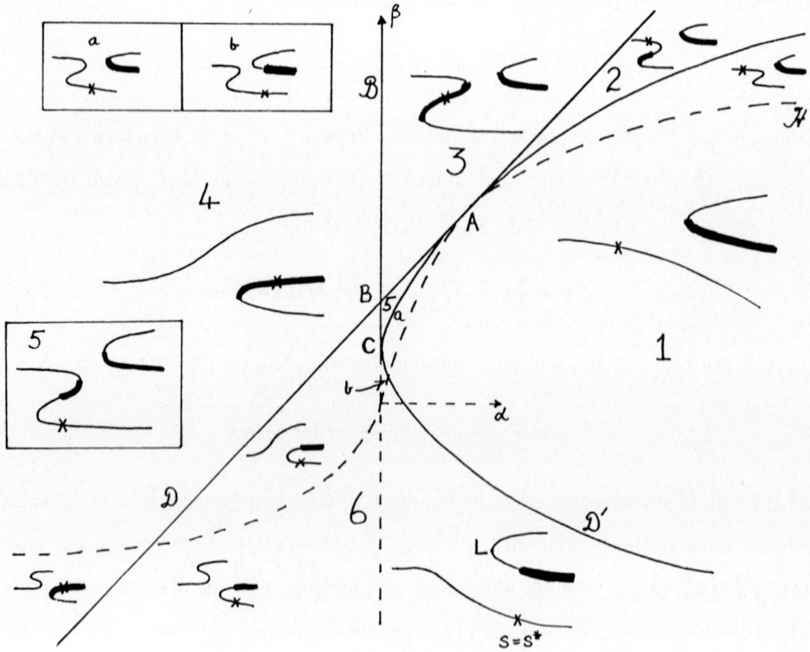

Fig. 6. Bifurcation diagrams for (4.1), $\varepsilon = A_{21} < 0$.

In between the curves $\mathcal{B}, \mathcal{H}, \mathcal{D}, \mathcal{D}'$, we draw bifurcation diagrams, indicating admissible portions of the curves with heavy lines. We also show the point $d = 0$ as a cross on the diagrams. At values of s for which there are three equilibria, the outer two are saddle points. Since U_- is always a saddle point in this local analysis, a cross on an outer branch indicates an undercompressive shock. By way of illustration, we show in Fig. 7 how the phase plane changes in case 6. The heavy line in Fig. 6, case 6, indicates fast compressive shocks, with corresponding saddle to node connections. As s decreases, the connecting trajectory approaches the saddle point U^*, until at $s = s^*$, there is a trajectory from U_- to U^*. For $s < s^*$, the unstable manifold from U_- misses U^*, and can therefore be connected to no other equilibrium (in this local picture). This is why the heavy line in case 6, Fig. 6 stops at $s = s^*$, rather than continuing all the way to the limit point L, as it would in the classical theory. From Fig. 6 we see that bifurcation diagrams on either side of the hysteresis curve $\mathcal{H}$ contain the same information concerning admissible shocks. The hysteresis curve $\mathcal{H}$ can therefore be disregarded. Similarly, the portion of the bifurcation locus $\mathcal{B}$ below the point C plays no role. Note that Figs. 4,7 are concerned with global behavior of the vector field in different small ranges of s. In Fig. 4, the equilibria that are close are near U_-, and U_+ is always a saddle. In Fig. 7, three U_+ equilibria are close, whereas U_- is far away, and is always a saddle point.

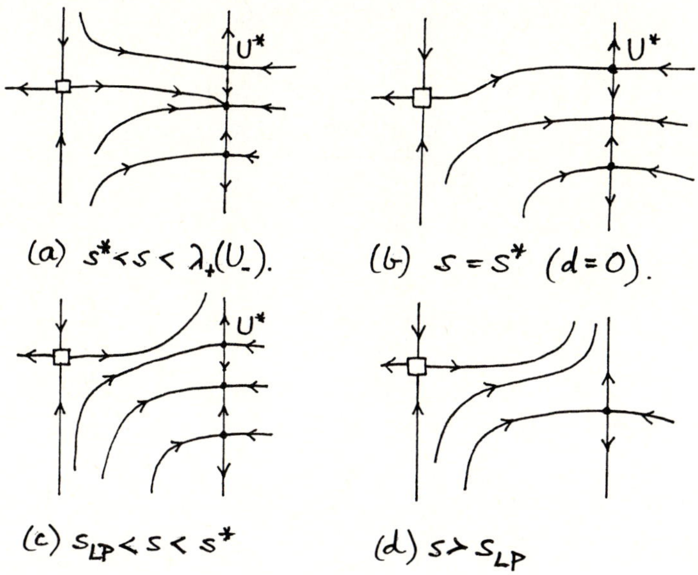

Fig. 7. Phase plane for (4.1), Case 6 of Fig. 6.
($s = s_{LP}$ at limit point.)

Now consider b, ϵ fixed and small. Then as v_- varies, (α, β) lies on a curve that cuts the curves $\mathcal{B}, \mathcal{H}, \mathcal{D}, \mathcal{D}'$ transversally. To first order, this curve is a straight line. The location of the straight line in Fig. 7 defines the precise sequence of bifurcation diagrams as v_- varies. There are three special cases, where the line includes the points A, B, or C in Fig. 7. Thus, Fig. 8, showing values of (b, ϵ) corresponding to the special cases, provides a subclassification of the equations. In terms of bifurcation theory, while the shock speed s is a distinguished bifurcation parameter, v_- is a distinguished perturbation parameter, and b, ϵ belong to a third level in the hierarchy of parameters.

5. Solution of Riemann problems. Here we solve the Riemann problem for system (4.1),(4.2) for $a = -1, b$ near zero, with $A_{21} = \epsilon < 0$ fixed near zero, and (u_L, v_L) near the median M_1 with $u_L < 0$. We also primarily restrict (u_R, v_R) to the neighborhood of values studied in §4, for which we understand the role of undercompressive shocks.

The pictures in Fig. 9, representing the solution of Riemann problems, have been deduced from our knowledge of how Riemann problem solutions are constructed, as explained in [4], and from the detailed understanding of admissibility of shocks gained from sections 3,4. As found at the end of §4, there are four qualitatively different solutions of the Riemann problem in the neighborhood under consideration, depending on the location of b and ϵ in Fig. 8.

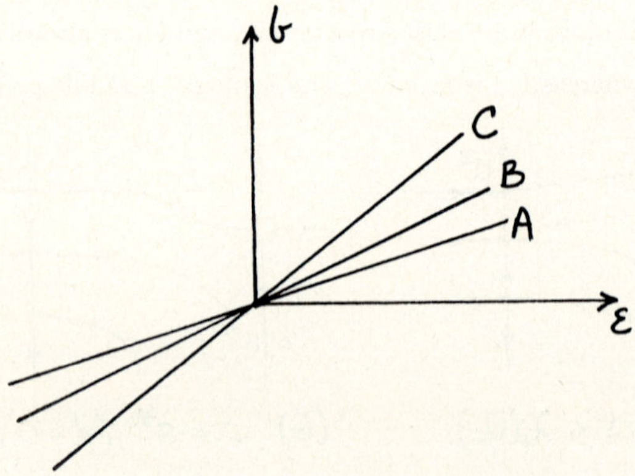

Fig. 8. Classification in the b, ε plane.

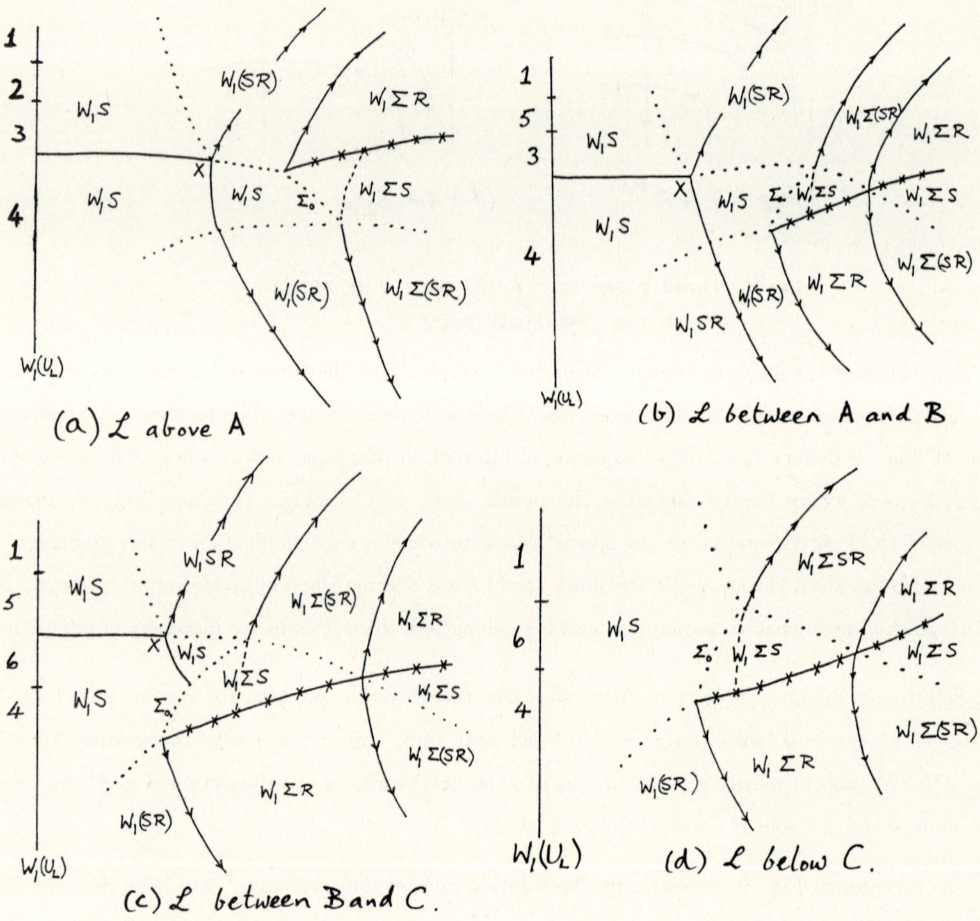

Fig. 9. Solution of Riemann problems for (4.1), ε < 0.

In Fig. 9, we have suppressed the precise location of U_L near the median, by showing the slow wave curve $W_1(U_L)$ in generalized form, without its detailed subdivision into shocks, rarefactions and rarefaction shocks. Thus for each U_L and U_R in Fig. 9, the solution of the Riemann problem has a slow wave adjacent to U_L in the (x,t)-plane, with an intermediate state on the right of the wave given by an appropriate point on $W_1(U_L)$. Each (u_-, v_-) on $W_1(U_L)$ corresponds to a point on a line $\mathcal{L}$ cutting across the curves of Fig. 7. The location of $\mathcal{L}$ with respect to the points A, B, C depends on the location of b in Fig. 8 (recall that we keep $\epsilon < 0$ fixed, and consider all possible small values of b). The numbers marked on $W_1(U_L)$ indicate ranges of v_- corresponding to the regions shown in Fig. 6.

The point X in Fig. 9 corresponds to a secondary bifurcation. The portion of the Hugoniot locus corresponding to admissible fast shocks and including X is shown. The curve Σ_o plays a special role. For each point $\tilde{U}$ on Σ_o, there is a U_- on $W_1(U_L)$, and a point U_+ on Σ such that $U_- \to U_+$ is a shock with speed s and there is a trajectory from U_+ to $\tilde{U}$ with speed s. Thus $\tilde{U}$ is the end point of a fast shock curve for U_- and the end point of a fast shock curve for U_+. If U_R lies on one side of Σ_o, then the Riemann problem solution involves a slow wave and a single fast shock. If U_R lies on the other side of Σ_o, then the solution involves a slow wave, an undercompressive shock and a fast shock with almost the same speed. As U_R approaches Σ_o, the shock speeds approach s. This construction is similar to that of overcompressive shocks [4]. The dashed line in Fig. 9 represents one edge of the region of values of U_R where the second construction applies (the dashed line is in fact a fast shock curve).

References

(1) C. C. Chicone, *Quadratic gradients on the plane are generically Morse-Smale*, J. Differential Equations, 33 (1979), 159-166.

(2) M. Golubitsky and D. G. Schaeffer, Singularities and Groups in Bifurcation Theory, Vol. 1, Springer, 1985.

(3) D. G. Schaeffer and M. Shearer, *The classification of* 2×2 *systems of nonstrictly hyperbolic conservation laws, with application to oil recovery*, Comm. Pure Applied Math., 4 (1987), 141-178.

(4) D. G. Schaeffer and M. Shearer, *Riemann problems for nonstrictly hyperbolic* 2×2 *systems of conservation laws*, Trans. A.M.S., 304 (1987), 267-306.

(5) S. Schecter, Interaction of equilibrium and heteroclinic bifurcation for planar vector fields, Preprint.

(6) S. Schecter and M. Shearer, Travelling wave solutions of viscous perturbations of nonstrictly hyperbolic conservation laws. In preparation.

(7) M. Shearer, D. G. Schaeffer, D. Marchesin and P. J. Paes-Leme, *Solution of the Riemann problem for a prototype* 2×2 *system of nonstrictly hyperbolic conservation laws*, Arch. Rational Mech. Anal., 97 (1987), 299-320.

(8) M. Shearer, Solution of the Riemann problem for case I nonstrictly hyperbolic systems of conservation laws. In preparation.

(9) Z. Tang and T.C.T. Ting, *Wave curves for the Riemann problem of plane waves in isotropic elastic solids*, International J. Engineering Science, 25 (1987), 1343-1381.

The first author was supported in part by National Science Foundation grant No. DMS-8701348, and by Air Force Office of Scientific Research grant No. AFOSR-87-0283.

THE VISCOSITY-CAPILLARITY APPROACH TO PHASE TRANSITIONS

Marshall Slemrod
Center for the Mathematical Sciences
University of Wisconsin - Madison
Madison, Wisconsin 53705

1. Goals of the hydrodynamic theory.

The goals of the hydrodynamic theory of phase transitions are to understand the changes of mass, momentum, and energy in fluids undergoing change of phase. Understanding may be at different levels: qualitative and asymptotic analysis of solutions; existence, uniqueness theorems; numerical algorithms and machine computation; analysis of special flows and comparison with experiment. In this note I will attempt to illustrate some progress in one special area of the above program namely study of the initial value problem for the one dimensional isothermal motion of a van der Waals fluid. The reason for this choice of problem is that it represents possibly the simplest hydrodynamic problem amenable to mathematical analysis. Understanding it will hopefully teach us how to analyse more complicated and probably more physically relevant problems where other phenomena have been incorporated e.g. thermodynamics, chemical reactions, etc.

2. Basic equations.

We wish to integrate the initial value problem (in Lagrangian coordinates)

$$u_t + p(w)_x = 0 \tag{2.1}$$

$$w_t - u_x = 0 \tag{2.2}$$

where u denotes velocity, w denotes specific volume, p denotes the pressure. We consider the case of a van der Waals like material where $p' < 0$ on $(-\infty,\alpha)$, (β,∞) and $p' > 0$ on (α,β), p a smooth function. We note the system (2.1), (2.2) is hyperbolic on $(-\infty,\alpha) \cup (\beta,\infty)$ and elliptic on the spinodal region (α,β).

For initial data we take piecewise constant data

$$w = \begin{cases} w_- & x < 0 \\ w_+ & x > 0 \end{cases}, \quad u = \begin{cases} u_- & x < 0 \\ u_+ & x > 0 \end{cases}. \tag{2.3}$$

Hence (2.1) - (2.3) is a Riemann initial value problem for a van der Waals like material.

We suppose furthermore that the data is in the physically realizable hyperbolic regimes, say

$w_- < \alpha, \quad w_+ > \beta$

where typically $w < \alpha$ represents the liquid phase and $w > \beta$ represents the vapor phase.

3. Data consistent with the Rankine-Hugoniot conditions.

The Rankine-Hugoniot jump conditions for (2.1) - (2.2) are

$$-s[u] + [p] = 0 \quad , \tag{3.1}$$

$$-s[w] - [u] = 0 \quad . \tag{3.2}$$

These conditions provide the consistency conditions that must be satisfied across a propagating curve $x(t)$ where u, w suffer jump discontinuities, $s = \frac{dx}{dt}$.

If the inital data satisfy the Rankine-Hugoniot jump conditions then

$$w = \begin{matrix} w_- & x < st \\ w_+ & x > st \end{matrix} \qquad u = \begin{matrix} u_- & x < st \\ u_+ & x > st \end{matrix} \tag{3.3}$$

is indeed a solution to the Riemann problem. While this result is in some sense trivial it is not without interest. This is because (3.3) represents a weak not strong solution of (2.1 - 2.3). As weak solutions of conservation laws are notoriously non-unique it seems reasonable to introduce an <u>admissibility criterion for weak solutions</u>.

A criterion I have suggested is the <u>viscosity-capillarity admissibility criterion</u>. It is based on the philosophy that the macroscopic solutions should be limits of microstructural solutions obtained for a system which accounts for both viscosity, capillarity, and thermal conductivity. In our isothermal set up we would say a weak solution u, w of (2.1) - (2.3) is <u>admissible</u> if it is the limit as $\varepsilon \to 0+$ of solutions of the microstructural system.

$$u_t + p(w)_x = \varepsilon u_{xx} - \varepsilon^2 A w_{xxx} \tag{3.4}$$

$$w_t - u_x = 0 \quad . \tag{3.5}$$

Here ε denotes the viscosity, $\varepsilon^2 A$ - the capillarity coefficient.

The idea of using such a structural approach to static phase transitions dates back to van der Waals [1]. Vis-a-vis dynamics similar ideas have been put foward by Cahn & Hilliard [2]. I believe however that the idea of using both viscosity and capillarity in studying hydrodynamic phase transitions orginated with my paper [3]. Other work in this area which extended this idea may be found in the papers of Shearer [4], [5], Hagan & Slemrod [6], Grinfeld [7], Hagan & Serrin [8].

Returning to our Riemann problem we note one way to check whether (3.3) is admissible according to the viscosity-capillarity criterion is to take traveling wave solutions of (3.4) - (3.5),

$$u = u(\xi),\ w = w(\xi),\ \xi = \frac{x-st}{\varepsilon}\ .$$

Substitution into (3.1) - (3.2) shows that solvability of the boundary value problem

$$-s(u-u_-) + p(w) - p(w_-) = u' - Aw'' \tag{3.6}$$

$$-s(w-w_{-\_} - (u-u_-) = 0 \tag{3.7}$$

$$u(-\infty) = u_-,\ w(-\infty) = w_-,\ u(+\infty) = u_+,\ w(+\infty) = w_+ \tag{3.8}$$

will provide verification that (3.3) is an admissible weak solution.

For example, where $u_- = u_+ = s = 0$, (3.6) becomes

$$-Aw'' = p(w) - p(w_-)\ . \tag{3.9}$$

An easy calculation (just multiply by w' and integrate from $-\infty$ to ∞) shows (3.9) with boundary conditions $w(-\infty) = w_-$, $w(+\infty) = w_+$ has a solution only if

$$\int_{w_-}^{w_+} (p(w) - p(w_-))dw = 0 \tag{3.10}$$

is satisfied. This is the Maxwell equal area rule. Explicit integration of (3.9) shows (3.10) is also sufficient for solvability.

While this is only a static result and our main interest is dynamics its seems pleasant to know our approach delivers the classical result for equilibrium co-existence of phases in a van der Waals like material. More details on genuinely dynamic problems may be found in [3], [5], [6], [7], [8], [9].

Of course it is possible to make theories that exclude viscosity or capillarity. This would model materials where perhaps both were present but one dominates the other. For example a viscosity dominant theory would model phase changes in a visco-elastic material. In this regard see the papers of Pego [10], Andrews & Ball [11], Shearer [4]. Capillarity dominant theories may be found in solid state physics where viscosity is negligible; see for example Maugin [12] and work referred to there.

4. Data not consistent with Rankine-Hugpniot jump conditions.

For general piecewise data we face a quandary. How to extend the notion of viscosity-capillarity limits and still make use of the special nature of the Riemann problems? In [13] I have suggested a way which extends earlier ideas of Kalasnikov [14], Tupciev [15], Dafermos [16], [17], and Dafermos & DiPerna [18].

The idea contains two pieces. First note that the substitutions

$$v = u - D_2 w_x$$

$$\begin{matrix} D_1 \\ D_2 \end{matrix} = \frac{\varepsilon}{2} \mp \frac{\varepsilon}{2}(1-4A)^{1/2}$$

brings (2.1) - (2.2) into diagonal form

$$v_t + p(w)_x = D_1 v_{xx} \quad , \tag{4.1}$$

$$w_t - v_x = D_2 w_{xx} \tag{4.2}$$

when $0 < A \leq 1/4$. In particular the choice $A = 1/4$, $\varepsilon = 2\bar{\varepsilon}$ would say that admissible solutions of (2.1) - (2.2) would limits of solutions of (4.1) - (4.2) as $D_1 = D_2 = \bar{\varepsilon} \to 0+$.

In fact (4.1) - (4.2) is a common form for many "artificial" viscosity arguments. Notice that it arises in a very "unartificial" way, i.e. as a direct change of variables from the balance laws of mass and momentum which included both viscosity <u>and</u> capillarity. (For D_1 and D_2 to both be non-zero both viscosity and capillarity must be present.)

The next step is due to Kalasnikov, Tupciev, and independently Dafermos. Simply stated it notes that solutions of Riemann problem for systems of conservation laws

$$U_t + F(U)_x = 0 \quad , \tag{4.3}$$

$$U = U_- \quad x < 0, \quad U = U_+ \quad x > 0 \tag{4.4}$$

$U \in \mathbb{R}^n$, $F : \mathbb{R}^n \to \mathbb{R}^n$ would be expected to be of the form $U = U(\xi)$, $\xi = \frac{x}{t}$ since (4.3), (4.4) is invariant to coordinate changes $x \to \alpha x$, $t \to \alpha t$, $\alpha > 0$ (a constant). So $U = U(\xi)$ is the expected macrostructural similarity form. Thus it would be tempting to such similarity solutions of artificially viscous systems

$$U_+ + F(U)_x = \varepsilon U_{xx} \tag{4.5}$$

and let $\varepsilon \to 0+$ to recover a weak solution of the Riemann problem. Unfortunately the εU_{xx} term in (4.5) does not preserve the desired scaling when $x \to \alpha t$ but $\varepsilon t U_{xx}$ does! This motivates the following problem. Solve

$$u_t + p(w)_x = \varepsilon t u_{xx} \quad , \tag{4.6}$$

$$w_t - u_x = \varepsilon t w_{xx} \quad , \tag{4.7}$$

for a similarity solution $u^\varepsilon(\xi)$, $w^\varepsilon(\xi)$ which satisfies boundary conditions

$$u(-\infty) = u_-, \; w(-\infty) = w_-, \; u(+\infty) = u_+, \; w(+\infty) = w_+ \tag{4.8}$$

and then take limits as $\varepsilon \to 0+$ of u^ε, w^ε to recover a solution of the Riemann problem. Tautologically we take this solution to be admissible since it is a viscosity-capillarity limit. However it is nice to know that Dafermos [17] has shown in the hyperbolic case of 2×2 systems of conservation laws that such limits do indeed satisfy the usual Lax admissibility conditions.

Substitution of $u(\xi)$, $w(\xi)$ into (4.6), (4.7) leads to the non-autonomous boundary value problem P_ε

$$-\xi u' + p(w)' = \varepsilon u'' \tag{4.9}$$

$$-\xi w' - u' = \varepsilon w'' \tag{4.10}$$

with boundary conditions (4.8).

In [13] the following has been proven:

<u>Result</u>. P_ε has solutions for which $w' > 0$ ($w' < 0$) in the region $\alpha \leq w \leq \beta$ when $w_- < w_+$ ($w_+ < w_-$).

The issue of passage to the limits as $\varepsilon \to 0$ is more difficult. The main results of [13] are the following when $w_- < w_+$.

<u>Assume</u> $p(w) \to +\infty$ as $w \to \infty$, $p(w) \to -\infty$ as $w \to -\infty$, $p''(w) > 0$ if $w \leq \alpha$, $p''(w) < 0$ if $w \geq \beta$.

<u>Case 1</u>. u^ε has a sequence of minima $u^\varepsilon(\tau^\varepsilon)$ for which $u^\varepsilon(\tau^\varepsilon) \to -\infty$, $\tau^\varepsilon \to 0$, as $\varepsilon \to 0+$.

<u>Case 2</u>. u^ε has no such sequence.

<u>Result</u>. In Case 2 $(u^\varepsilon, w^\varepsilon) \to (u,w)$ a.e. on $(-\infty,\infty)$ where u, w is a weak solution to (2.1) - (2.3).

In Case 1 $(u^\varepsilon, w^\varepsilon) \to (u,w)$ a.e. on $(-\infty,\infty)$ where u, w is a weak solution of (2.1) - (2.3) on $(-\infty,0) \cup (0,\infty)$. The pair (u,w) will be a solution of the Riemann problem if the pressure p equilibriates across the stagnant phase boundary $\xi = 0$, i.e.

$$\lim_{\xi \to 0-} p(w(\xi)) = \lim_{\xi \to 0+} p(w(\xi)) \quad .$$

In fact we know $-(p(\beta) - p(\alpha)) \leq \lim_{\xi \to 0-} - p(w(\xi)) + \lim_{\xi \to 0+} p(w(\xi)) \leq 0$ holds in general. This shows us that as we approach the hyperbolic case where $p(\beta) = p(\alpha)$ (the critical isotherm) any possible errors made in the limiting procedure disappear. It seems an unfortunate bit of irony that the stagnant phase boundary that was so easily recovered in Section 3 via the viscosity-capillarity limit causes the trouble when trying to analyse the general Riemann problem. This is the price we pay for introducing the εt term.

Acknowledgement

This research was sponsored in part by the Air Force Office of Scientific Research, Air Force Systems Comman, USAF Contrac/Grant No. AFOSR 87-0315. The U.S. Government's right to retain a non exclusive royalty free license in and to copyright this paper for governmental purposes is acknowledged.

References

[1] J. D. van der Waals, The thermodynamic theory of capillarity under the hypothesis of a continuous variation of density (in Dutch), Verh. Konink. Acad. Wetensch., Amsterdam (Sec 1), Vol. No. 8 (1893).

[2] J. W. Cahn and J. E. Hilliard, Free energy of a nonuniform system, I: Interfacial free energy. J. Chem. Phys., 28, 258-267 (1958).

[3] M. Slemrod, Admissibility criteria for propagating phase boundaries in a van der Waals fluid, Arch. Rational Mech. Anal., 81, 301-315 (1983).

[4] M. Shearer, The Riemann problem for a clas of conservation laws of mixed type, J. Differential Equations, 46, 426-443 (1982).

[5] M. Shearer, Nonuniqueness of admissible solutions of the Riemann initial value problem for a system of conservation laws of mixed type, Arch. Rational Mech. Anal., 93, 45-59 (1986).

[6] R. Hagan and M. Slemrod, The viscosity-capillarity criterion for shocks and phas transitions, Arch. Rational Mech. Anal., 83, 333-361 (1984).

[7] M. Grinfeld, Topological techniques in the dynamics of phas transitions, Ph.D. Thesis, Rensselaer Polytechnic Institute (1986).

[8] R. Hagan and J. Serrin, One dimensional shock layers in Korteweg fluids, in Phase Transformations and Material Instabilities, ed. M. E. Gurtin, Academic Press, New York (1984).

[9] M. Shearer, Riemann problem for a van der Waals fluid, to appear Quarterly of Applied Mathematics.

[10] R. Pego, Phase transitions in one dimensional nonlinear viscoelasticity: Admissibility and stability, Arch. Rational Mech. Anal., 353-394 (1987).

[11] G. Andrews and J. M. Ball, Asymptotic behavior of changes of phase in one dimensional nonlinear viscoelasticity, J. Differential Equations 6, 71-86 (1982).

[12] G. Maugin, Continuum Mechanics of Electro-magnetic Solids, North-Holland (Amsterdam), 1988.

[13] M. Slemrod, A limiting "viscosity" approach to the Riemann problem for materials exhibiting change of phase, to appear Arch. Rational Mech. Anal.

[14] A. S. Kalasnikov, Construction of generalized solutions of quasilinear equations of first order without convexity conditions as limits of solutions of parabolic equations with a small parameter, Dokl. Akad. Nauk SSR 127, 27-30 (1959) (Russian).

[15] V. A. Tupciev, The asymptotic behavior of solution of the Cauchy problem for the equation $\varepsilon^2 tu_{xx} = u_t + [\varphi(u)]_x$ that degenerates for $\xi = 0$ into the problem of the decay of an arbitrary discontinuity for the case of a rarefaction wave. Zh. Vychisl. Mat. i Fiz. 12, 770-775 (1972). English translation in USSR, Comput. Math. and Phys. 12.

[16] C. M. Dafermos, Solution of the Riemann problem for a class of hyperbolic systems of conservation laws by the viscosity method, Arch. Rational Mech. Anal., 52, 1-9 (1973).

[17] C. M. Dafermos, Structure of solutions of the Riemann problem for hyperbolic systems of conservation laws, Arch. Rational Mech. Anal., 53, 203-217 (1974).

[18] C. M. Dafermos and R. DiPerna, The Riemann problem for certain classes of hyperbolic systems of conservation laws, J. Differential Equations, 90-114 (1976).

A VERSION OF THE FUNDAMENTAL THEOREM FOR YOUNG MEASURES

J.M. Ball
Department of Mathematics, Heriot-Watt University,
Riccarton, Edinburgh EH14 4AS, Scotland.

1. Introduction.

Let $\Omega \subset \mathbb{R}^n$ be measurable and let $z^{(j)}:\Omega \longrightarrow \mathbb{R}^m$ be a given sequence of functions. The fundamental theorem concerning Young measures asserts that under appropriate hypotheses there exists a subsequence $z^{(\mu)}$ of $z^{(j)}$ and a family of probability measures (ν_x) on $\mathbb{R}^m$ (the *Young measure*) such that the weak limit of $f(z^{(\mu)})$ is given by the expectation

$$<\nu_x, f> = \int_{\mathbb{R}^m} f(\lambda)\, d\nu_x(\lambda) \quad \text{a.e. } x \in \Omega, \tag{1}$$

for any continuous function $f:\mathbb{R}^m \longrightarrow \mathbb{R}$. The purpose of this note is to give a version of the theorem that is convenient for some applications to nonlinear partial differential equations and variational problems of mechanics, together with a reasonably self-contained proof. Although sharper than some statements in the literature as regards the hypotheses on the $z^{(j)}$ and f, the form of the theorem given here is not essentially new. In particular, E.J. Balder has shown me how it may be regarded as a consequence of a very general lower semicontinuity theorem he has proved in [3 Theorem 2.1]. Nevertheless, I hope that the statement given here may be of some use.

The Young measure (ν_x) can intuitively be thought of as giving the limiting probability distribution as $\mu \longrightarrow \infty$ of the values of $z^{(\mu)}$ near x. To be more precise, suppose that Ω is open and that $x \in \Omega$. Denote by $B(x,\delta)$ the open ball with centre x and radius $\delta > 0$. Keeping x, μ and δ fixed, let $\nu^{(\mu)}_{x,\delta}$ be the probability distribution of the values of $z^{(\mu)}(y)$ as y is chosen uniformly at random from $B(x,\delta)$. Then it is shown below that

$$\nu_x = \lim_{\delta\to 0} \lim_{\mu\to\infty} \nu^{(\mu)}_{x,\delta}, \tag{2}$$

almost everywhere, where the convergence is weak* in the sense of measures.

The Young measure was introduced by L.C. Young (see [39,40]) as a means of treating problems of the calculus of variations for which there does not exist a minimizer in a classical sense. Many applications and developments to the calculus of variations and optimal control theory have been made by MacShane [28], Gamkrelidze [24] and others. More abstract ideas in the same spirit are the integral currents of Federer & Fleming [23] and varifolds (see, for example, Allard [1]). The Young measure was developed as a tool for analysing nonlinear partial differential equations by Tartar [36], who suggested how it could help to prove existence theorems for nonlinear hyperbolic systems, a programme he carried out for a single hyperbolic equation. These ideas were then significantly developed by DiPerna [18] to prove the existence of solutions for a system of two hyperbolic equations in one space dimension. Further results along these lines have been given by, for example, Dafermos [15], DiPerna [18], Rascle [31], Roytburd & Slemrod [32], Schonbek [33] and Serre [34]. Applications of the Young measure to variational problems of continuum mechanics have been made by Ball & Knowles [9], Ball & James [7,8], Chipot & Kinderlehrer [14], and Kinderlehrer [27]. In the last four papers cited, the application is to the description of the microstructure of crystals. (The order in which the limits are taken in (2) corresponds to the way in which the microstructure is experimentally observed; namely, a small region of diameter δ of the crystal is examined microscopically, but this region is typically larger than the length scale μ^{-1} of the microstructure.)

The method used here to prove the fundamental theorem delivers (ν_x) directly via duality rather than by disintegration of a measure on a product space; the principal idea can be found in Castaing & Valadier [13], Warga [38,39], Balder [3,4], and recent descriptions appear in Capuzzo Dolcetta & Ishii [12], and Slemrod & Roytburd [35]. An alternative direct construction is suggested in Tartar [37 pp268-9]. For the more usual method see Berliocchi & Lasry [11], Tartar [36] and Balakrishnan [2]. A Young measure corresponding to a bounded sequence in L^p capable of detecting some concentrations as well as oscillations has been introduced by DiPerna & Majda [19,20] as a tool for studying vortex dynamics.

2. The fundamental theorem for Young measures.

Our aim is to prove the following result:

Theorem

Let $\Omega \subset \mathbb{R}^n$ be Lebesgue measurable, let $K \subset \mathbb{R}^m$ be closed, and let $z^{(j)}: \Omega \longrightarrow \mathbb{R}^m$, $j = 1,2,\ldots$, be a sequence of Lebesgue measurable functions satisfying $z^{(j)}(\cdot) \longrightarrow K$ in measure as $j \longrightarrow \infty$, i.e. given any open neighbourhood U of K in $\mathbb{R}^m$

$$\lim_{j\to\infty} \text{meas}\,\{x \in \Omega : z^{(j)}(x) \notin U\} = 0.$$

Then there exists a subsequence $z^{(\mu)}$ of $z^{(j)}$ and a family (ν_x), $x \in \Omega$, of positive measures on $\mathbb{R}^m$, depending measurably on x, such that

(i) $\qquad \|\nu_x\|_M := \int_{\mathbb{R}^m} d\nu_x \le 1 \quad$ for a.e. $x \in \Omega$,

(ii) $\qquad \text{supp}\, \nu_x \subset K$ for a.e. $x \in \Omega$, and

(iii) $\qquad f(z^{(\mu)}) \xrightarrow{\;*\;} \langle \nu_x, f\rangle = \int_{\mathbb{R}^m} f(\lambda)\, d\nu_x(\lambda)$

in $L^\infty(\Omega)$ for each continuous function $f : \mathbb{R}^m \longrightarrow \mathbb{R}$ satisfying

$\lim_{|\lambda|\to\infty} f(\lambda) = 0.$

Suppose further that $\{z^{(\mu)}\}$ satisfies the boundedness condition

$$\lim_{k\to\infty} \sup_{\mu} \text{meas}\, \{x \in \Omega \cap B_R : |z^{(\mu)}(x)| \ge k\} = 0, \tag{3}$$

for every $R > 0$, where $B_R = B(0,R)$. Then $\|\nu_x\|_M = 1$ for a.e. $x \in \Omega$ (i.e. ν_x is a probability measure), and given any measurable subset A of Ω

$$f(z^{(\mu)}) \longrightarrow \langle \nu_x, f\rangle \qquad \text{in } L^1(A) \tag{4}$$

for any continuous function $f : \mathbb{R}^m \longrightarrow \mathbb{R}$ such that $\{f(z^{(\mu)})\}$ is sequentially weakly relatively compact in $L^1(A)$.

Remarks

1. The condition (3) is very weak, and is equivalent to the following: given any $R > 0$ there exists a continuous nondecreasing function $g_R : [0,\infty) \longrightarrow \mathbb{R}$, with $\lim_{t\to\infty} g_R(t) = \infty$, such that

$$\sup_{\mu} \int_{\Omega\cap B_R} g_R(|z^{(\mu)}(x)|)\, dx < \infty. \tag{5}$$

In fact suppose that (5) holds. Then, since g_R is nondecreasing,

$$\sup_{\mu} \text{meas}\, \{x \in \Omega \cap B_R : |z^{(\mu)}(x)| \ge t\}\cdot g_R(t) \le \sup_{\mu} \int_{\Omega\cap B_R} g_R(|z^{(\mu)}(x)|)\, dx.$$

Since $\lim_{t\to\infty} g_R(t) = \infty$, we obtain (3).

Conversely, if (3) holds, we may choose $0 < t_j < t_{j+1}$, $j=1,2,\dots$, so that

$$\sup_{\mu} \text{meas}\, \{x \in \Omega\cap B_R : |z^{(\mu)}(x)| \ge t_j\} \le j^{-3},$$

and let

$$\bar{g}_R(t) = \begin{cases} 0 & \text{if } t \in [0, t_1), \\ j & \text{if } t \in [t_j, t_{j+1}). \end{cases}$$

Then

$$\sup_{\mu} \int_{\Omega\cap B_R} \bar{g}_R(|z^{(\mu)}(x)|)\, dx = \sup_{\mu} \sum_{j=1}^{\infty} j\, \text{meas}\, \{x \in \Omega \cap B_R : t_{j+1} > |z^{(\mu)}(x)| > t_j\}$$

$$\le \sum_{j=1}^{\infty} j^{-2} < \infty .$$

Choosing a suitable continuous $g_R \le \bar{g}_R$ we thus obtain (5).

Conditions similar to (5) are used by Berliocchi & Lasry [11 Proposition 5], and by Balder [3 Section 2] (who calls it 'tightness'; see also Balder [5]).

An application of the theorem to the case when the $z^{(j)}$ are bounded in $L^1(\Omega;\mathbb{R}^m)$ (i.e. with $g_R(t) = t$) appears in Ball & Murat [10].

2. If the functions $z^{(j)}$ are uniformly bounded in $L^\infty(\Omega;\mathbb{R}^m)$ then the functions $f(z^{(j)})$ are uniformly bounded in $L^\infty(\Omega)$ for any continuous $f:\mathbb{R}^m \longrightarrow \mathbb{R}$. Hence by the theorem there is a family of probability measures (ν_x) and a subsequence $z^{(\mu)}$ such that

$$f(z^{(\mu)}) \xrightarrow{\ *\ } \langle \nu_x, f \rangle \quad \text{in } L^\infty(\Omega)$$

for all such f. In this way we recover the form of the theorem given by Tartar [36]. If Ω is bounded and if the $z^{(j)}$ are uniformly bounded in $L^p(\Omega;\mathbb{R}^m)$ for some p, $1 < p < \infty$, then we obtain from the theorem the existence of a family of probability measures (ν_x) and a subsequence $z^{(\mu)}$ such that

$$f(z^{(\mu)}) \longrightarrow \langle \nu_x, f \rangle \quad \text{in } L^r(\Omega) \tag{6}$$

for any continuous $f : \mathbb{R}^m \longrightarrow \mathbb{R}$ satisfying

$$|f(\lambda)| \le \mathit{const.}\ (1 + |\lambda|^q), \qquad \lambda \in \mathbb{R}^m, \tag{7}$$

where $q > 0$ and $1 < r < p/q$ (see Schonbek [33]).

3. If A is bounded, the condition that $\{f(z^{(\mu)})\}$ be sequentially weakly relatively compact in $L^1(A)$ is satisfied if and only if

$$\sup_\mu \int_A \psi(|f(z^{(\mu)})|)\, dx < \infty$$

for some continuous function $\psi : [0,\infty) \longrightarrow \mathbb{R}$ with $\lim_{\lambda\to\infty} \psi(\lambda)/\lambda = \infty$ (de la Vallée Poussin's criterion; cf. MacShane [29], Dellacherie & Meyer [16]).

4. As explained in the introduction, the Young measure (ν_x) can be thought of as the limiting probability distribution of the values of $z^{(\mu)}$ near the point x. In fact if Ω is open and $x \in \Omega$ then for $\delta > 0$ sufficiently small the open ball $B(x,\delta)$ with centre x and radius δ is contained in Ω, and the formula

$$\langle \nu^{(\mu)}_{x,\delta}, f \rangle = ⨍_{B(x,\delta)} f(z^{(\mu)}(y))\, dy \tag{8}$$

defines a continuous linear form on continuous functions $f : \mathbb{R}^m \longrightarrow \mathbb{R}$ of compact support. (Here and below $⨍_E (\cdot)\, dx$ denotes $(\text{meas } E)^{-1}\int_E (\cdot)\, dx$.) The Radon measure $\nu^{(\mu)}_{x,\delta}$ is a probability measure giving the distribution of the values of $z^{(\mu)}$ in $B(x,\delta)$, and can be written

$$\nu^{(\mu)}_{x,\delta} = ⨍_{B(x,\delta)} \delta_{z^{(\mu)}(y)}\, dy, \tag{9}$$

where δ_a denotes the Dirac mass at $a \in \mathbb{R}^m$. As $\mu \longrightarrow \infty$, $\nu^{(\mu)}_{x,\delta} \overset{*}{\longrightarrow} \nu_{x,\delta}$ in the sense of measures, where by the theorem

$$<\nu_{x,\delta}, f> \; = \; ⨍_{B(x,\delta)} <\nu_y, f> dy \,, \tag{10}$$

that is,

$$\nu_{x,\delta} \; = \; ⨍_{B(x,\delta)} \nu_y \; dy \,. \tag{11}$$

By Lebesgue's differentiation theorem, for any fixed f

$$<\nu_{x,\delta}, f> \;\longrightarrow\; <\nu_x, f> \qquad \text{as } \delta \longrightarrow 0 \text{ for a.e. } x \in \Omega.$$

Choosing a countable dense set of functions f we deduce easily that $\nu_{x,\delta} \overset{*}{\longrightarrow} \nu_x$ in the sense of measures as $\delta \longrightarrow 0$ for a.e. $x \in \Omega$. The number $1 - \|\nu_x\|_M$ represents the limiting proportion of the points in $B(x,\delta)$ at which $z^{(\mu)}$ becomes unbounded.

Proof of the theorem

We denote by $C_0(\mathbb{R}^m)$ the Banach space of continuous functions $f : \mathbb{R}^m \longrightarrow \mathbb{R}$ satisfying $\lim_{|\lambda|\to\infty} f(\lambda) = 0$, with the norm $\|f\|_{C^0} = \sup_{\lambda\in\mathbb{R}^m} |f(\lambda)|$. A well known form of the Riesz representation theorem (Hewitt & Stromberg [25 p364]) asserts that there is a isometric isomorphism between the dual space $C_0(\mathbb{R}^m)^*$ of $C_0(\mathbb{R}^m)$ and the Banach space $M(\mathbb{R}^m)$ of bounded Radon measures on $\mathbb{R}^m$ obtained by associating with each $\nu \in M(\mathbb{R}^m)$ the linear form $f \longmapsto \int_{\mathbb{R}^m} f(\lambda)\, d\nu$ on $C_0(\mathbb{R}^m)$. The norm on $M(\mathbb{R}^m)$ is given by $\|\nu\|_M = \int_{\mathbb{R}^m} d|\nu|$. We associate with $z^{(j)}$the mapping $\nu^{(j)}: \Omega \longrightarrow M(\mathbb{R}^m)$ defined by

$$\nu^{(j)}(x) \; = \; \delta_{z^{(j)}(x)} . \tag{12}$$

For each j, $\nu^{(j)}$ belongs to the space $L^\infty_w(\Omega; M(\mathbb{R}^m))$ of equivalence classes of weak* measurable mappings $\mu : \Omega \longrightarrow M(\mathbb{R}^m)$ that are essentially bounded, i.e.

$$\|\mu\|_{\infty,M} := \operatorname*{ess\,sup}_{x\in\Omega} \|\mu(x)\|_M < \infty \,. \tag{13}$$

(We say that μ is weak* measurable if $<\mu(x), f>$ is measurable with respect to x for every $f \in C_0(\mathbb{R}^m)$.) Under the norm $\|\cdot\|_{\infty,M}$, $L^\infty_w(\Omega; M(\mathbb{R}^m))$ is a Banach space. Since $C_0(\mathbb{R}^m)$ is separable, there is an isometric isomorphism between the dual space of $L^1(\Omega; C_0(\mathbb{R}^m))$ and $L^\infty_w(\Omega; M(\mathbb{R}^m))$ obtained by associating with each $\mu \in L^\infty_w(\Omega; M(\mathbb{R}^m))$ the linear form

$$\psi \longmapsto \int_\Omega <\mu(x), \psi(x,\cdot)> dx \tag{14}$$

on $L^1(\Omega; C_0(\mathbb{R}^m))$ (see Edwards [22 p588], A. & C. Ionescu Tulcea [26 p93], Meyer [30 p244]). But

$$\|\nu^{(j)}\|_{\infty,M} = 1 \quad \text{for all } j.$$

Since $C_0(\mathbb{R}^m)$ is separable, so is $L^1(\Omega; C_0(\mathbb{R}^m))$, and hence (cf. Dunford &

Schwartz [21 pp 424-426]) there exists a subsequence $\nu^{(\mu)}$ of $\nu^{(j)}$ and an element $\nu = (\nu_x)$ of $L^\infty_w(\Omega; M(\mathbb{R}^m))$ such that $\nu^{(\mu)} \xrightarrow{*} \nu$ in $L^\infty_w(\Omega; M(\mathbb{R}^m))$. By (14) this implies that

$$\int_\Omega \psi(x, z^{(\mu)}(x))\, dx \longrightarrow \int_\Omega \langle \nu_x, \psi(x,\cdot)\rangle\, dx \tag{15}$$

as $\mu \longrightarrow \infty$ for every $\psi \in L^1(\Omega; C_0(\mathbb{R}^m))$. In particular, taking $\psi(x,\lambda) = \phi(x)f(\lambda)$, where $\phi \in L^1(\Omega)$ and $f \in C_0(\mathbb{R}^m)$, we have that

$$f(z^{(\mu)}) \xrightarrow{*} \langle \nu_x, f\rangle \quad \text{in } L^\infty(\Omega) \tag{16}$$

for every $f \in C_0(\mathbb{R}^m)$, which is (*iii*). By weak* lower semicontinuity of the norm, $\|\nu\|_{\infty,M} \le 1$, which is (*i*). To prove (*ii*), assume that $K \ne \mathbb{R}^m$ and let $f \in C^K_0(\mathbb{R}^m) := \{g \in C_0(\mathbb{R}^m)\ ;\ g|_K = 0\}$. Then, taking $U = \{z \in \mathbb{R}^m : |f(z)| < \varepsilon\}$, it follows from the hypothesis that $z^{(j)}(\cdot) \longrightarrow K$ in measure that $f(z^{(\mu)}(\cdot)) \longrightarrow 0$ in measure. Since f is bounded, we deduce that

$$\begin{aligned}\int_\Omega \phi(x)\, \langle \nu_x, f\rangle\, dx &= \lim_{\mu\to\infty}\int_\Omega \phi(x)\, f(z^{(\mu)}(x))\, dx \\ &= 0\end{aligned}$$

for every $\phi \in L^1(\Omega)$. Since $C_0(\mathbb{R}^m)$ is separable so is $C^K_0(\mathbb{R}^m)$, and it follows that for a.e. $x \in \Omega$ we have $\langle \nu_x, f\rangle = 0$ for all $f \in C^K_0(\mathbb{R}^m)$, i.e. supp $\nu_x \subset K$. Since $\nu^{(j)}(x) \ge 0$ a.e. we deduce similarly that $\nu_x \ge 0$ a.e..

Now suppose that (3) holds. Define $\vartheta^k \in C_0(\mathbb{R}^m)$ by

$$\vartheta^k(\lambda) = \begin{cases} 1 & \text{for } |\lambda| \le k, \\ 1 + k - |\lambda| & \text{for } k \le |\lambda| \le k+1, \\ 0 & \text{for } |\lambda| \ge k+1. \end{cases} \tag{17}$$

Then if $E \subset \Omega$ is bounded and measurable

$$\begin{aligned}\lim_{\mu\to\infty} ⨍_E \vartheta^{(k)}(z^{(\mu)}(x))\, dx &= ⨍_E \langle \nu_x, \vartheta^{(k)}\rangle\, dx \\ &\le ⨍_E \|\nu_x\|_M\, dx .\end{aligned} \tag{18}$$

But

$$0 \le ⨍_E \left[1 - \vartheta^{(k)}(z^{(\mu)}(x))\right] dx \le \frac{\text{meas}\{x \in E : |z^{(\mu)}(x)| \ge k\}}{\text{meas } E},$$

so that letting $k \longrightarrow \infty$ we get from (3) and (18) that

$$1 \le ⨍_E \|\nu_x\|_M\, dx . \tag{19}$$

Since $\|\nu_x\|_M \le 1$ a.e. and E is arbitrary, (19) implies that $\|\nu_x\|_M = 1$ a.e..

Suppose further that $f : \mathbb{R}^m \longrightarrow \mathbb{R}$ is continuous and that $f(z^{(\mu)})$ is sequentially weakly relatively compact in $L^1(\Omega)$. Let $f^+ = \max(f, 0)$, $f^- = \max(-f, 0)$, so that $f = f^+ - f^-$. By the Dunford-Pettis theorem (Dunford & Schwartz [21 p292]) $f^+(z^{(\mu)})$ and $f^-(z^{(\mu)})$ are both sequentially weakly relatively compact in $L^1(\Omega)$. Hence to prove (4) we can and will suppose that

$f \geq 0$. It also clearly suffices to prove (4) for the case when A is bounded and $f(z^{(\mu)}) \longrightarrow \chi$, say, in $L^1(A)$. Define $f^k \in C_0(\mathbb{R}^m)$ by $f^k = \vartheta^k f$, where ϑ^k is given by (17). Let $\phi \in L^\infty(A)$. We claim that

$$\int_A \phi\, f^k(z^{(\mu)})\, dx \longrightarrow \int_A \phi\, f(z^{(\mu)})\, dx \tag{20}$$

as $k \longrightarrow \infty$, uniformly in μ. In fact

$$\left| \int_A \phi\, (f^k(z^{(\mu)}) - f(z^{(\mu)}))\, dx \right|$$

$$\leq const. \int_{\{x\in A:\, |z^{(\mu)}(x)| \geq k\}} f(z^{(\mu)})\, dx\,, \tag{21}$$

and, given $\varepsilon > 0$, by the Dunford-Pettis theorem there exists an $M > 0$ such that

$$\sup_\mu \int_{\{x\in A:\, f(z^{(\mu)}(x)) \geq M\}} f(z^{(\mu)})\, dx \leq \varepsilon\,.$$

Hence by (3)

$$\int_{\{x\in A:\, |z^{(\mu)}(x)| \geq k\}} f(z^{(\mu)})\, dx \leq \varepsilon + M \text{ meas } \{x\in A : |z^{(\mu)}(x)| \geq k\}$$

$$\leq 2\varepsilon\,,$$

for all μ if k is sufficiently large, which together with (21) gives (20). Since also by (*iii*)

$$\lim_{\mu\to\infty} \int_A \phi\, f^k(z^{(\mu)})\, dx = \int_A \phi\, \langle \nu_x, f^k \rangle\, dx \tag{22}$$

it follows that

$$\lim_{k\to\infty} \int_A \phi\, \langle \nu_x, f^k \rangle\, dx = \int_A \phi\, \chi\, dx\,. \tag{23}$$

Choosing $\phi \geq 0$ and noting that f^k is increasing, we deduce from the monotone convergence theorem that $\langle \nu_x, f \rangle = \chi$ a.e. in A as required.

□

Remarks on the proof

1. The proof makes contact in several places with that of Balder [3 Theorem 2.1].

2. The reader is warned that the mappings $\nu^{(j)}$ do not in general belong to the space $L^\infty(\Omega;\, M(\mathbb{R}^m))$ of essentially bounded strongly measurable mappings from Ω to $M(\mathbb{R}^m)$ even when the $z^{(j)}$ are smooth. This space cannot be identified with the dual of $L^1(\Omega; C_0(\mathbb{R}^m))$ via (14).

3. The same argument as in the proof shows that under the hypothesis (3), for any measurable $A \subset \Omega$,

$$f(\cdot, z^{(\mu)}) \longrightarrow \langle \nu_x, f(x,\cdot) \rangle \text{ in } L^1(A) \tag{24}$$

for every Carathéodory function $f : A \times \mathbb{R}^m \longrightarrow \mathbb{R}$ such that $\{f(\cdot, z^{(\mu)})\}$ is sequentially weakly relatively compact in $L^1(A)$.

Acknowledgements

This research was carried out during a visit to the Institute for Mathematics and its Applications at the University of Minnesota in 1987. I am grateful to M.Luskin for having made this visit possible and pleasant, to J.L.Ericksen and R.D.James for having prompted the investigation, and to D.Kinderlehrer for discussions. I also wish to thank E.J.Balder for his comments on an earlier version. The paper was completed while I visited the Laboratoire d'Analyse Numérique, Université Pierre et Marie Curie, Paris with the support of C.N.R.S..

References

[1] W.K.Allard, "*On the first variation of a varifold*", Annals of Math.,**95** (1972) 417-491.

[2] A.V.Balakrishnan, '*Applied Functional Analysis*', Springer, 1976.

[3] E.J.Balder, "*A general approach to lower semicontinuity and lower closure in optimal control theory*", SIAM J. Control and Optimization, **22** (1984) 570-598.

[4] E.J.Balder, "*Generalized equilibrium results for games with incomplete information*", Mathematics of Operations Research, 13 (1988) 265-276.

[5] E.J.Balder, "*Fatou's lemma in infinite dimensions*", J. Math. Anal. Appl., **136** (1988).

[6] J.M.Ball, "*Material instabilities and the calculus of variations*", Proc. conference on '*Phase transformations and material instabilities in solids*', Mathematics Research Center, University of Wisconsin, Academic Press, Publication No. **52**, (1984) 1-20.

[7] J.M.Ball and R D James, "*Fine phase mixtures as minimizers of energy*", Arch. Rat. Mech. Anal.,**100** (1987) 13-52.

[8] J.M.Ball and R.D.James, "*Proposed experimental tests of a theory of fine microstructure, and the two-well problem*", to appear.

[9] J.M.Ball and G.Knowles, unpublished work summarised in [6].

[10] J.M.Ball and F.Murat, "*Remarks on Chacon's biting lemma*", to appear.

[11] H.Berliocchi and J.M.Lasry, "*Intégrandes normales et mesures paramétrées en calcul des variations*", Bull. Soc. Math. France, **101** (1973) 129-184.

[12] I.Capuzzo Dolcetta and H.Ishii, "*Approximate solutions of the Bellman equation of deterministic control theory*", Appl. Math. Optim., **11** (1984) 161-181.

[13] C.Castaing and M.Valadier, '*Convex Analysis and Measurable Multi--functions*', Springer Lecture Notes in Mathematics, Vol.**580**, 1977.

[14] M.Chipot and D.Kinderlehrer, "*Equilibrium configurations of crystals*", Arch. Rat. Mech. Anal., **102** (1988) 237-278.

[15] C.M.Dafermos, "*Solutions in L^∞ for a conservation law with memory*", in '*Analyse Mathématique et Applications*' Gauthier-Villars, (1988).

[16] C.Dellacherie and P-A.Meyer, '*Probabilités et Potentiel*',Hermann, 1975.

[17] R.J.DiPerna, "*Convergence of approximate solutions to conservation laws*", Arch. Rat. Mech. Anal., **82** (1983) 27-70.

[18] R.J.DiPerna, "*Convergence of the viscosity method for isentropic gas dynamics*", Comm. Math. Phys. **91** (1983) 1- 30.

[19] R.J.DiPerna and A.J.Majda, "*Oscillations and concentrations in weak solutions of the incompressible fluid equations*", Comm. Math. Phys., **108** (1987) 667-689.

[20] R.J.DiPerna and A.J.Majda, "*Concentrations in regularizations for 2-D incompressible flow*", Comm. Pure Appl. Math., **40** (1987) 301-345.

[21] N.Dunford and J.T.Schwartz, '*Linear Operators*',*Part* I, Interscience, 1967.

[22] R.E.Edwards, '*Functional Analysis*', Holt, Rinehart and Winston, 1965.

[23] H.Federer and W.H.Fleming, "*Normal and integral currents*", Annals of Math., **72** (1960) 458-520.

[24] R.V.Gamkrelidze, "*On sliding optimal states*", Dokl. Akad. Nauk. SSSR **143** (1962) 1243-1245 = Soviet Math. Doklady, **3** (1962) 559-561.

[25] E.Hewitt and K.Stromberg, '*Real and Abstract Analysis*', Springer, 1965.

[26] A.& C. Ionescu Tulcea, '*Topics in the Theory of Lifting*', Springer, New York, 1969.

[27] D.Kinderlehrer, "*Remarks about equilibrium configurations of crystals*", in '*Material Instabilities in Continuum Mechanics*', ed. J.M.Ball, Oxford University Press, 1988, pp.217-241.

[28] E.J.MacShane, "*Generalized curves*", Duke Math. J., **6** (1940) 513-536.

[29] E.J.MacShane, '*Integration*', Princeton Univ. Press, 1947.

[30] P-A.Meyer, '*Probability and Potentials*', Blaisdell, Waltham, 1966.

[31] M.Rascle, "*Un résultat de 'compacité par compensation' à coefficients variables. Application a l'élasticité non linéaire*", C.R. Acad. Sci. Paris, **302** (1986) 311-314.

[32] V.Roytburd & M.Slemrod, "*An application of the method of compensated compactness to a problem in phase transitions*", in '*Material Instabilities in Continuum Mechanics*', ed. J.M.Ball, Oxford University Press, 1988, pp.427-463.

[33] M.E.Schonbek, "*Convergence of solutions to nonlinear dispersive equations*", Comm. in Partial Diff. Equations, **7** (1982) 959-1000.

[34] D.Serre, "*La compacité par compensation pour les systèmes hyperboliques non linéaires à une dimension d'espace*", J. Math. Pure et Appl., **65** (1987) 423-468.

[35] M.Slemrod and V.Roytburd, "*Measure-valued solutions to a problem in dynamic phase transitions*", in '*Contemporary Mathematics*', Vol.**50**, Amer. Math. Soc., 1987.

[36] L.Tartar, "*Compensated compactness and applications to partial differential equations*", in '*Nonlinear Analysis and Mechanics*', *Heriot-Watt Symposium*, Vol IV, Pitman Research Notes in Mathematics, 1979, pp.136-192.

[37] L.Tartar, "*The compensated compactness method applied to systems of conservation laws*", in '*Systems of Nonlinear Partial Differential Equations*', ed.J.M.Ball, NATO ASI Series, Vol. C111, Reidel, 1982, pp.263-285.

[38] J.Warga, "*Relaxed variational problems*", J. Math. Anal. Appl., **4** (1962) 111-128.

[39] J.Warga, '*Optimal Control of Differential and Functional Equations*', Academic Press, 1972.

[40] L.C.Young, "*Generalized curves and the existence of an attained absolute minimum in the calculus of variations*", Comptes Rendus de la Société des Sciences et des Lettres de Varsovie, classe III, **30** (1937) 212-234.

[41] L.C.Young, '*Lectures on the Calculus of Variations and Optimal Control Theory*', Saunders, 1969 (reprinted by Chelsea, 1980).

VERY SLOW PHASE SEPARATION IN ONE DIMENSION

J. Carr
Department of Mathematics
Heriot-Watt University
Riccarton, Edinburgh EH14 4AS

and

R. L. Pego
Department of Mathematics
University of Michigan
Ann Arbor, MI 48109

ABSTRACT

We consider a scalar Landau-Ginzburg equation with nonconserved order parameter: $u_t = \epsilon^2 u_{xx} - F'(u)$ for $0 \leqslant x \leqslant 1$ with Neumann boundary conditions, where F is a double well free energy density with equal minima at $u = \pm 1$. Given typical mixed-phase initial data, over a short time domains form where u is near ± 1 alternately. Collapse of domains typically occurs on a time scale which is exponentially large, proportional to $\exp(A\ell/\epsilon)$ where $A^2 = F''(\pm 1)$ and ℓ is the domain length. From this we give a (nonrigorous) scaling law for the coarsening process in one dimension. Physically, the motion of an interface between domains is driven by the tiny potential jump across it. For our rigorous results, we take a dynamical systems approach. In geometric terms, we find thin channels which attract the flow and along which certain slow variables dominate. Proofs will appear elsewhere.

1. In this paper we describe some new rigorous results [1] concerning the extremely slow evolution of patterns in solutions of the equation

$$u_t = \epsilon^2 u_{xx} - f(u) \tag{1.1}$$

for $0 \leqslant x \leqslant 1$, $t > 0$, subject to Neumann boundary conditions

$$u_x(0,t) = 0, \quad u_x(1,t) = 0, \quad \text{for} \quad t > 0 \ .$$

Here f satisfies $f(\pm 1) = 0$, $f'(\pm 1) > 0$, $\int_{-1}^{1} f(u)du = 0$, so we may write $f(u) = F'(u)$ where $F(\pm 1) = F'(\pm 1) = 0$. We assume that $F(u) > 0$ for $-1 < u < 1$. For simplicity, we assume here (but not in [1]) that f is odd and $\limsup_{u \to \pm\infty} f(u)/u > 0$. Prototypical is $f(u) = u^3 - u$.

Equation (1.1) is a Landau-Ginzburg equation with nonconserved order parameter and is widely used in the kinetics of phase transitions. The parameter ϵ is proportional to a correlation length and F is the free energy density, having equal minima at $u = \pm 1$.

Analytically, (1.1) is a dissipative system. A Lyapunov function is the total energy

$$I(u) = \int_0^1 \left[\frac{1}{2}\epsilon^2 u_x^2 + F(u)\right] dx$$

for which

$$\frac{d}{dt} I(u) = -\int_0^1 u_t^2 dx < 0 \quad \text{unless } u \text{ is a stationary solution.}$$

Using this, it may be shown [2] that for any solution $\lim_{t\to\infty} u(x,t) = u_\infty(x)$ exists and that u_∞ is a stationary solution of (1.1). Furthermore, the only stable stationary solutions of (1.1) are constant in space, such as $u \equiv 1$ and $u \equiv -1$. Typical solutions may be expected then to converge to one of these "patternless" stable states as $t \to \infty$.

2. Observing solutions computed numerically, one's expectations become modified. Figure 1 plots a solution of (1.1) computed on the interval $[-1,1]$ with $\epsilon = .1$, $f(u) = u^3 - u$, and initial data $u(x,0) = \arctan((x + .2)/\epsilon)$. Computation was performed in double precision using the NAG routine DØ3PAF, which uses finite differences in space, Gear's method in time. The solution may well seem to have converged to a stable steady state at $t = 10$. By $t = 10^7$, nothing has visibly changed. The solution is in a "metastable" state. A relatively sudden final convergence to $u \equiv +1$ occurs at about $t = 1.14 \times 10^8$.

We might reason loosely as follows. If ϵ is small, for a fixed time taking $\epsilon = 0$ is a good approximation. (See [4] for a rigorous result.) This time is of the order of $t = 2$ for the computation above. A pattern of transition layers develops as diffusion balances the layer steepening induced by the ODE $u_t = -f(u)$. These layers should behave roughly like the standing wave solution $U(x)$ of (1.1) defined for $-\infty < x < \infty$, satisfying

$$\epsilon^2 U_{xx} - f(U) = 0, \quad U(-\infty) = -1, \quad U(\infty) = 1, \quad U(0) = 0 \quad . \tag{2.1}$$

For $f(u) = \frac{1}{2}(u^3 - u)$, $U(x) = \tanh(x/2\epsilon)$. So layers tend not to move, although generally they must.

3. We aim to describe the slow evolution of solutions of (1.1) with N transition layers by obtaining equations of motion for N variables $h_j(t)$ describing the layer locations. When $f(u) = \frac{1}{2}(u^3 - u)$ and $N = 1$, the equation obtained for $h = h_1$ is

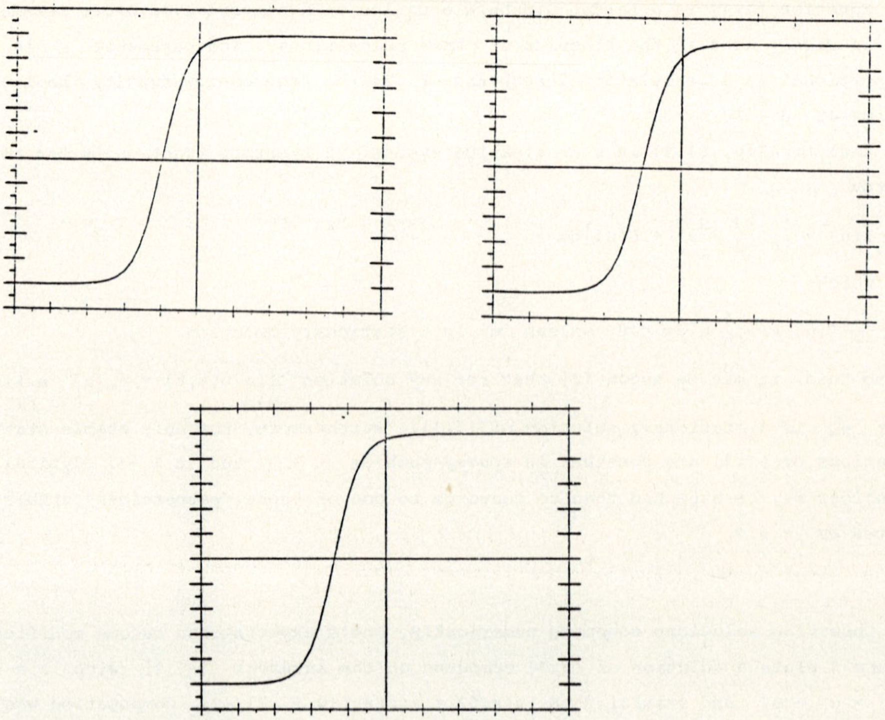

Figure 1.

A solution of (1.1) computed at (a) $t = 10$, (b) $t = 10^7$

(c) $t = 10^8$.

$$\dot{h} = -12\epsilon(e^{-2h/\epsilon} - e^{-2(1-h)/\epsilon}) \quad . \tag{3.1}$$

This equation can be derived formally by a method due to J. Neu [3]. To give insight, we sketch this derivation. Let $U(x)$ be the solution of (2.1), so that $U(x) = \tanh(x/2\epsilon)$. A solution of (1.1) with one transition layer at $x = h$ is assumed to be approximately described near the layer by

$$u(x,t) \approx U(x - h(t)),\ u_x \approx U'(x - h(t)),\ u_t \approx -\dot{h}U'(x - h(t)) \quad .$$

For any solution u,

$$u_x u_t = \left(\frac{1}{2}\,\epsilon^2 u_x^2 - F(u)\right)_x = -q_x \quad . \tag{3.2}$$

Away from the layer, $u_t \approx 0$, so that by (3.2),

$$\frac{1}{2}\,\epsilon^2 u_x^2 - F(u) = -q \approx \begin{cases} -q_- & \text{for } x < h \\ -q_+ & \text{for } x > h \end{cases} \quad . \tag{3.3}$$

Using these approximations in (3.2) and integrating gives

$$+\dot{h}\int_0^1 U'(x-h)^2dx \approx q_+ - q_- \quad . \tag{3.4}$$

From (2.1), $(\epsilon U')^2 = \frac{1}{4}(1-u^2)^2$ so the integral in (3.4) is given approximately by

$$\frac{1}{2\epsilon}\int_{-1}^{1}(1-u^2)du = \frac{2}{3\epsilon} \quad .$$

Away from the layer the solution is approximated by linearizing about $u = \pm 1$ so that by (3.3) we get

$$\frac{1}{2}(\epsilon u_x)^2 - \frac{1}{2}(u \pm 1)^2 \approx q_{\mp} \quad . \tag{3.5}$$

This equation can be solved to give solutions valid in the regions $x < h$ and $x > h$, satisfying the boundary conditions. Matching these approximations with the approximation $u \approx U(x-h)$ gives

$$q_- \approx 8e^{-2h/\epsilon}, \quad q_+ \approx 8e^{-2(1-h)/\epsilon} \quad .$$

From these, equation (3.1) follows. Equations of motion for N layers may be obtained in a similar way. Physically, the mechanism driving the motion of layers is seen from (3.4) to be differences in the values of the "potential"

$$q = F(u) - \frac{1}{2}\epsilon^2 u_x^2 \quad .$$

4. Our goal is to rigorously justify equation (3.1) and its generalization for N layers and to explain the stability of the metastable states associated with a given configuration of N layers. We only outline the main ideas and results from this investigation, which are based on a "dynamical systems" approach, viewing evolving solutions in terms of the geometry of function space. Full details will appear in [1].

Fix N. Given a configuration $h = (h_1,\ldots,h_N)$ of N layer positions, well separated with $0 < h_1 < \ldots < h_N < 1$, we associate a function $u^h(x)$ which approximates a metastable state. (See Figure 2. We assume $u^h(0) < 0$ in light of the symmetry $u \to -u$.) Let

$$\mathcal{L}(u) = \epsilon^2 u_{xx} - f(u) \quad . \tag{4.1}$$

We construct u^h so that $\mathcal{L}(u^h)$ is very small; in fact $\mathcal{L}(u^h)(x) = 0$ except near the layers, as suggested by numerical results and the formal calculations in §3. The precise definition of u^h appears in the appendix. The admissible configurations h lie in a set defined by

$$\Omega = \{h : h_{j+1} - h_j > \epsilon H, \quad j = 0,\ldots,N\} \tag{4.2}$$

where H is some constant, and $h_0 = -h_1$, $h_{N+1} - 1 = 1 - h_N$. All the results to be described in this paper require H to be sufficiently large, and we will not mention this again. There is no other restriction on ϵ other than the obvious $0 < \epsilon < 1/NH$.

The set of the states u^h form an N-dimensional manifold $M = \{u^h : h \in \Omega\}$. The idea is that slowly evolving solutions with N layers will be close to M and

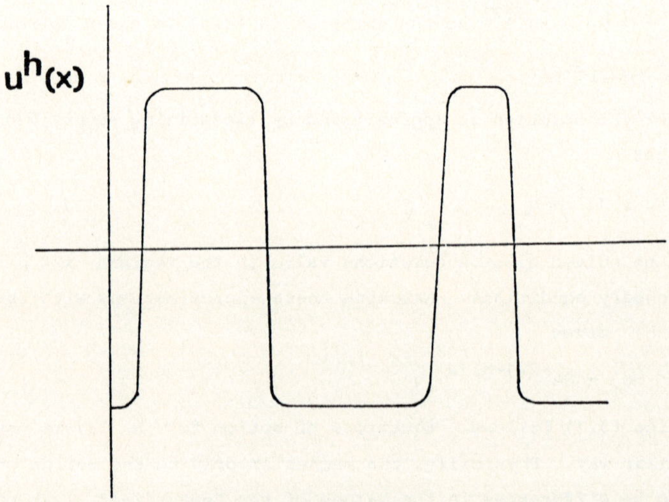

Figure 2
A typical $u^h(x)$ for $N = 4$.

that M will determine the singular part of the flow. To study this we introduce co-ordinates relative to M. The tangent space to M at u^h is spanned by $\{\partial u^h/\partial h_j : j = 1,\dots,N\}$; for technical convenience we work with approximate tangent vectors τ_j^h, defined in the appendix in terms of u_x^h. The approximation is based on the identity $\frac{\partial}{\partial h} U(x - h) = -\frac{\partial}{\partial x} U(x - h)$. For a function $u(x)$ close to M there exist unique coordinates u^h, v such that $u = u^h + v$ where $u^h \in M$ and $\langle \tau_j^h, v\rangle = 0$, $j = 1,\dots,N$. $\langle \tau, v\rangle = \int_0^1 \tau v dx$ denotes the inner product in $L^2(0,1)$. Thus given a solution $u(x,t)$ of (1.1) near M we may write

$$u(x,t) = u^{h(t)}(x) + v(x,t) \quad . \tag{4.3}$$

Using this representation we can write (1.1) in the form

$$\begin{aligned} \dot{h}_j &= \epsilon G_j(h) + g_j(h,v), \quad j = 1,\dots,N \\ v_t &= -L^h v + \sigma(h,v) \quad . \end{aligned} \tag{4.4}$$

Here L^h is the linearization of $-\mathcal{L}$ about u^h and is given by

$$L^h v = -\epsilon^2 v_{xx} + f'(u^h)v \quad . \tag{4.5}$$

The functions G_j have the form

$$G_j(h) = S[q(h_{j+1} - h_j) - q(h_j - h_{j-1})] \tag{4.6}$$

where $S^{-1} = \int_{-1}^{1} \sqrt{2F(u)}du$. The function $q(\ell)$ in (4.6) is the (constant) value of the potential $F(\phi) - \frac{1}{2}\epsilon^2\phi_x^2$ where ϕ is that steady state solution of (1.1) such that the distance between adjacent layers is ℓ (i.e. $\phi = \phi(x,\ell,\pm 1)$ as defined in the appendix). Near M, the terms g_j and σ are small in a sense to be detailed below. One of the basic results of [1] characterizes the leading term in $q(\ell)$ by a careful study of ϕ:

Theorem 1. For ℓ/ϵ sufficiently large,

$$q(\ell) = Ke^{-A\ell/\epsilon}(1 + O((\epsilon/\ell)e^{-A\ell/2\epsilon}))$$

where $A^2 = F''(+1)$ and $K = \frac{1}{2}A^2\kappa^2$ where

$$\kappa = 2\exp\left[\int_0^1\left[\frac{A}{\sqrt{2F(t-1)}} - \frac{1}{t}\right]dt\right] .$$

Corresponding formulae hold for derivatives of q with respect to ℓ.

5. In this section and the next we outline the rest of the main results of [1]. An important aspect of the study of (4.4) is the spectral analysis of L^h and its relation with the constraints on v. For insight, first consider the linear operator obtained by linearizing $\mathcal{L}(u)$ about $U(x)$ from (2.1):

$$w \to -\epsilon^2 w_{xx} + f'(U)w .$$

This operator has principal eigenvalue $\lambda = 0$ with eigenfunction $U'(x)$; this is connected with the fact that $\mathcal{L}(U(x-h)) = 0$ for all h. For the case at hand, the fact that $\frac{\partial}{\partial h_j}\mathcal{L}(u^h)$ is small for $j = 1,\ldots,N$ gives rise to exactly N tiny eigenvalues.

Theorem 2. There exist constants C_1, C_2 independent of ϵ, h such that if $h \in \Omega$, the eigenvalues $\lambda_1 < \lambda_2 < \cdots$ of L^h satisfy

$$|\lambda_j| \le C_1 e^{-A\ell^h/2\epsilon} \quad \text{for } j = 1,\ldots,N ,$$

$$\lambda_{N+1} \ge C_2 .$$

Here

$$\ell^h = \min\{h_j - h_{j-1}: j = 1,\ldots,N+1\} .$$

The eigenspace for the first N eigenvalues of L^h is almost parallel to the span of the approximate tangent vectors τ_j^h, so using Theorem 2 we can show that, normal to this span, L^h is coercive:

Theorem 3. There exists a constant C independent of ϵ, h such that if $h \in \Omega$

and $\langle\tau_j^h,v\rangle = 0$ for $j = 1,\dots,N$, then

$$\frac{C}{2}\,\epsilon\|v\|^2_{L^\infty} \le C\int_0^1 (\epsilon^2 v_x^2 + v^2)dx \le \langle L^h v,v\rangle \quad . \tag{5.1}$$

We use this result together with (4.4) to get the energy estimate below, valid for a solution of (1.1) decomposed in the form (4.3):

$$\left[\frac{d}{dt} + \frac{C}{2}\right]\left[\langle L^h v,v\rangle - \Gamma\epsilon|G|^2\right] \le 0 \quad . \tag{5.2}$$

Here $|G|^2 = \sum_{j=1}^{N} G_j(h)^2 \le Ce^{-2A\ell^h/\epsilon}$ from Theorem 1 and $\Gamma > 0$ is some constant.

The estimate (5.2) expresses the transverse stability of metastable states near M, in terms of an order 1 exponential rate of decay.

6. Using estimate (5.2) we can construct a channel for the flow. Define

$$Z = \{u : u = u^h + v,\ h \in \Omega,\ \langle\tau_j^h,v\rangle = 0 \text{ for } j = 1,\dots,N\ ,$$
$$\text{and } \langle L^h v,v\rangle \le \Gamma\epsilon|G(h)|^2\} \quad .$$

<u>Theorem 4</u>. Suppose $u(\cdot,t)$ lies in Z for $0 < t < T$ where T is maximal. Then
(i) $h(T) \in \partial\Omega$ if $T < \infty$,
(ii) $\|v\|_{L^\infty} \le \left[\frac{2}{\epsilon}\int_0^1(\epsilon^2 v_x^2 + v^2)\right]^{1/2} \le C|G(h)| \le C^2 e^{-A\ell^h/\epsilon}$ for $0 \le t \le T$.

That is, a solution cannot leave Z by violating the bounds on v, but only by having $h_j(T) - h_{j-1}(T) = \epsilon H$ for some j. That is, a pair of layers get too close together. The channel Z is exponentially thin in the v-direction.

Using these estimates for v, our main results characterize accurately the equation of motion for $h(t)$ from (4.3) when u lies in the channel Z. Below, $|g|^2 = \sum_{j=1}^{N} g_j^2$.

<u>Theorem 5</u>. If $u = u^h + v$ lies in Z, then

$$|g(h,v)| < C\epsilon|G(h)|e^{-A\ell^h/2\epsilon} \quad .$$

Thus the direction and magnitude of the rate of change of the configuration vector $h = (h_1,\dots,h_N)$ is determined by $G(h)$ in (4.6) to very high relative accuracy.

7. In this section we discuss the behaviour of solutions in the slow channel Z near M. Because of Theorem 5, this behaviour is determined by that in the system

$$\dot h_j = \epsilon G_j(h) = \epsilon S(q(h_{j+1} - h_j) - q(h_j - h_{j-1})) \quad . \tag{7.1}$$

This system is a gradient system, equivalent to

$$\dot{h}_j = -\frac{\partial W(h)}{\partial h_j} \tag{7.2}$$

where $W(h) = \epsilon S[\frac{1}{2} Q(\ell_0) + \sum_{j=1}^{N-1} Q(\ell_j) + \frac{1}{2} Q(\ell_N)]$

with $\ell_j = h_{j+1} - h_j$ for $j = 0,\ldots,N$ and $Q'(\ell) = q(\ell)$.

There is a unique steady state solution in Z, which we denote by $u^*(x)$ or $u^{h^*}(x)$, so that $G(h^*) = 0$. For this solution, the layer positions are evenly spaced: $h_j^* = (j - ½)/N$. The solution u^* is unstable in (1.1) and it may be shown that the linearization L^{h^*} has exactly N unstable (negative) eigenvalues.

All solutions of (1.1) in Z except $u = u^*$ <u>eventually</u> leave Z. This means that $h_j = h_{j-1} + \epsilon H$ for some $j = 1,\ldots,N+1$. That is, some pair of layers get close together or h_1 or h_N get close to their reflection in the boundary. A rigorous result concerning the persistence of the N-layers metastable states in Z is the following lower bound for the escape time:

<u>Theorem 6</u>. Suppose $u(x,t)$ is a solution of (1.1) lying in Z for $0 \leq t < T$ and T is maximal. (T is the escape time.) Suppose that at $t = 0$ we have $\ell^\circ \geq 3H\epsilon/2$ where

$$\ell^\circ = \min\{(h_{j+1} - h_j)|_{t=0} : j = 0,\ldots,N\} .$$

Then

$$T > B_- e^{A\ell^\circ/\epsilon}$$

where $B_-^{-1} = SA^3\kappa^2(1 + CHe^{-AH/2})$ for some constant C (independent of H for large H).

The potential function $q(\ell)$ is exquisitely sensitive to the parameters A, ℓ, and ϵ. When one domain length $\ell_i = h_{i+1} - h_i$ is much shorter than the others initially, one of the terms $q(\ell_i)$ in (4.4) strongly dominates. The solution may be reasonable described by a single ordinary differential equation for the shortest domain length $\ell = \ell_i$:

$$\dot{\ell} = -2\epsilon SKe^{-A\ell/\epsilon} . \tag{7.3}$$

The solution of (7.3) satisfies

$$\ell(t) = \ell(0) + \ell n(1 - \tau)^{\epsilon/A} \tag{7.4}$$

where $\tau = t/(SA^3\kappa^2)^{-1}e^{A\ell(0)/\epsilon}$. For the great proportion of time before it collapses to zero, $\ell(t)$ is virtually indistinguishable from $\ell(0)$.

A rigorous upper bound for escape time in this case is:

<u>Theorem 7</u>. Suppose $u(\cdot,0)$ lies in Z and for some L (sufficiently large),

$\ell_i(0) < \ell_j(0) - \epsilon L$, $j \neq i$ for some i. Then the escape time T satisfies

$$T < B_+ e^{A\ell_i(0)/\epsilon}$$

where $B_+^{-1} = SA^3\kappa^2(1 - CHe^{-AH/2} - Ce^{-AL})$ for some constant C (independent of H and L).

It is interesting to give some numbers to illustrate the sensitivity and size possible of escape times predicted from equation (7.4). First consider the plots in Figure 1, computed in $[-1,1]$ with $\epsilon = .1$, $f(u) = u^3 - u$. Taking $h(0) = -.2$, after rescaling and using equation (7.4) one predicts an escape time $T \simeq 1.4 \times 10^8$. Considering the sensitivity to the unknown correct value of $h(0)$, this is not a bad match for 1.14×10^8. Imagine, however, one computed in $[-\pi,\pi]$ with $\epsilon = .1$, $f(u) = 2(u^3 - u)$, a one-layer solution with $h(0) = -.2\pi$. After rescaling, equation (7.4) predicts an escape time $T > 4 \times 10^{41}$. Numerically, rounding errors would dominate u_t, even though the solution is nowhere near a steady state solution. Probably the best way to detect numerically that something was odd here would be to compute some eigenvalues of the linearization $w \to -\epsilon^2 w_{xx} + f'(u)w$. The small ones give a hint that there are some "slow variables" that should be considered.

8. To conclude, we give a nonrigorous description of the "coarsening" process which occurs as a solution which initially develops N layers progresses to a patternless stable steady state. Imagine that ϵ is very small and N is large. When the solution leaves the channel $Z_N = Z$, typically two layers approach each other as described in §7, then annihilate each other. The last stages of this process occur swiftly, on an order 1 time scale (see (5.2)), as the solution collapses into another channel Z_{N-2} corresponding to metastable states involving the $N - 2$ remaining layers ($N - 1$ if annihilation at the boundary occurs). One believes, though it is not proven, that the remaining $N - 2$ layer positions are essentially unaffected during this collapse. The solution cascades down into channels involving fewer and fewer dimensions, each time staying considerably longer than before, since longer domains are involved. Eventually the last layer collapses against the boundary and equilibrium is attained. It is possible that the solution could get stuck at some point and approach the unstable steady solution while collapsing onto some Z_j, but this is extremely unlikely. More likely it would never be near any steady solution until the end; it can fall into Z_j anywhere.

A more intrinsic geometric description of the coarsening process would involve recognition that the channel Z is actually a small neighbourhood of a large part of the unstable manifold of the unstable steady solution u^* in Z. Work in this direction is in progress.

Finally, we give a "scaling law" which should roughly describe the time dependence of a characteristic length during the coarsening process. It is based on equation (7.4) which approximately describes the collapse of a single domain. A domain with initial length ℓ collapses at approximately the time

$$T(\ell) = B_0 e^{A\ell/\epsilon}$$

where $B_0^{-1} - SA^3\kappa^2$. Thus we expect that after time t, no domain with length less than $L(t)$ will typically be observed, where the characteristic length $L(t)$ satisfies

$$L(t) = \epsilon A^{-1}\ell n\ t - \cdot A^{-1}\ell n\ B_0 \quad .$$

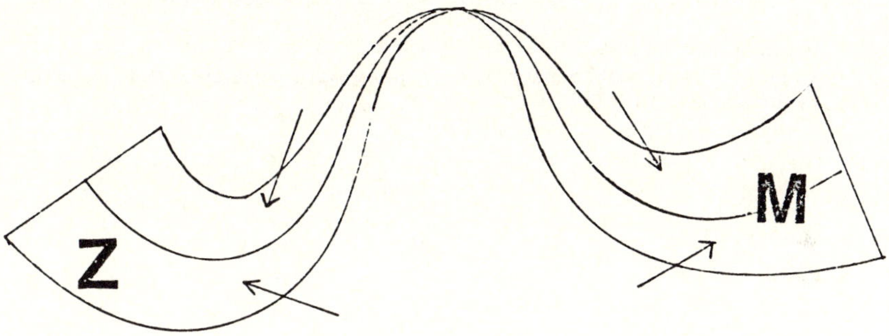

Figure 3
The channel Z.

APPENDIX

Let $\phi(x,\ell,{}^{+}1)$ be the solution of

$$\mathcal{L}(\varphi) = \epsilon^2\varphi_{xx} - f(\varphi) = 0, \quad \varphi = 0 \text{ at } x = -\ell/2 \text{ and } x = \ell/2 \quad ,$$

$$\pm\varphi(x,\ell,\pm 1) > 0 \text{ for } |x| < \ell/2 \quad .$$

Note that since we have assumed that f is odd, $\phi(x,\ell,+1) = -\phi(x,\ell,-1)$. Let $\chi : \mathbf{R} \to [0,1]$ be a smooth function with $\chi(x) = 0$ for $x \leqslant -1$ and $\chi(x) = 1$ for $x \geqslant 1$. Let $m_j = (h_{j-1} + h_j)/2$. Then the states u^h are defined by

$$u^h(x) = (1 - \chi((x - h_j)/\epsilon))\varphi(x - m_j, h_j - h_{j-1}, (-1)^j)$$

$$+ \chi((x - h)/\epsilon)\varphi(x - m_{j+1}, h_{j+1} - h_j, (-1)^{j+1}) \quad ,$$

for $x \in [m_j, m_{j+1}]$.

The approximate tangent vectors τ_j^h are defined by $\tau_j^h(x) = -\gamma^j(x)u_x^h(x)$ where

$$\gamma^j(x) = \chi((x - m_j - \epsilon)/\epsilon)[1 - \chi((x - m_{j+1} + \epsilon)/\epsilon)] \quad .$$

ACKNOWLEDGEMENTS

We are grateful for many enthusiastic discussions with J. Hale and G. Fusco who suggested the basic dynamical systems point of view which we have detailed and developed here and in [1]. Part of this work was carried out when the first author was visiting CRM, Universitat Autonoma de Barcelona. The second author was supported by a National Science Foundation postdoctoral fellowship and the SERC.

REFERENCES

[1] J. Carr & R. L. Pego, Metastable patterns in solutions of $u_t = \epsilon^2 u_{xx} - f(u)$, submitted.

[2] H. Matano, Asymptotic behaviour and stability of solutions of semilinear diffusion equations, Publ. Res. Inst. Math. Sci., Kyoto University, 15 (1979), 401-454.

[3] J. Neu, unpublished notes.

[4] H. Weinberger, On metastable patterns in parabolic systems, IMA preprint 164, University of Minnesota 1985.

LIST OF PARTICIPANTS

BALL John	Department of Mathematics, Heriot-Watt, University Riccarton EDINBURGH EH14 4AS SCOTLAND
BARANGER Jacques	Université Lyon I, LAN, Bt 101 69622 VILLEURBANNE Cedex FRANCE
BESTION Dominique	C E A , 85 X, 38041 GRENOBLE Cedex FRANCE
BILLIA Bernard	Laboratoire de Physique Cristalline, Faculté de St Jérôme, case 151 13397 MARSEILLE Cedex 13 FRANCE
BOLLEY Catherine	Ecole Nationale Supérieure de Mécanique, 1, Rue De La Noë 44072 NANTES Cedex 03 FRANCE
BOUCHITTE Guy	Laboratoire de Mécanique Rationnelle, Faculté des Sciences Saint Jérôme Rue Henri Poincaré; 13013 MARSEILLE Cedex FRANCE
BOULIGAND Yves	Centre de Cytologie Expérimentale, Ecole Pratique des Hautes Etudes, 67, Rue Maurice Günsbourg; 94200 IVRY-SUR-SEINE FRANCE
BREZIS Haim	Laboratoire d'Analyse Numérique, Université Paris VI,Tour 55-65 4, Place Jussieu; 75230 PARIS Cedex 05 FRANCE
CADET Serge	Laboratoire de Mécanique Théorique, Université Paris VI, Tour 66 4, Place Jussieu; 75230 PARIS Cedex 05 FRANCE
CALDERER Maria Carme	Department of Mathematics; George Mason University FAIRFAX, Virginia 22030 U.S.A.
COUDER Yves	Groupe de Physique du Solide, Ecole Normale Supérieure 24, Rue Lhomond; 75231 PARIS Cedex 05 FRANCE
COULLET Pierre	Physique Théorique, Université de Nice, Parc Valrose 06034 NICE Cedex FRANCE
FREMOND Michel	Service Informatique, L.C.P.C., 58, Bd Lefebvre 75732 PARIS Cedex 15 FRANCE
GLIMM James	Courant Institute of Mathematical Sciences, New York University 251, Mercer Street; NEW YORK, N.Y., 10012 U.S.A.
GOLDENFELD Nigel	Looms Laboratory of Physics, University of Illinois1110 West Green Street URBANA, I11. 61801 U.S.A.
GOUIN Henri	Faculté des Sciences et Techniques de Saint-Jérôme, Rue Poincaré 13397 MARSEILLE Cedex 13 FRANCE
GRINFELD Michael	Department of Mathematics Heriot-Watt University, Riccarton EDINBURGH EH 14 4AS SCOTLAND
GUILLOPE Colette	Département de Mathématiques, Bt 425, Université Paris-Sud 91405 ORSAY Cedex FRANCE
GURTIN Morton	Department of Mathematics Carnegie-Mellon University PITTSBURGH, PA, 15213-3890 U.S.A.
HENNENBERG Marcel	Laboratoire de Chimie-Physique II, Université libre de Bruxelles BRUXELLES BELGIQUE
JAMES Richard	Department of Aerospace Engineering and Mechanics, University of Minnesota; MINNEAPOLIS, MN 55455 U.S.A.

KINDERLEHRER David	Institute for Math. and its Application University of Minnesota Vincent Hall 514, Church street S.E. MINNEAPOLIS, MN 55455 U.S.A.
KLEMAN Maurice	Laboratoire de Physique du Solide, Université Paris-Sud 91405 ORSAY Cedex FRANCE
LENORMAND Roland	Etudes et Fabrication,Dowell Schlumberger, Z.I. Molina la Chazotte B.P. 90, 42000 SAINT ETIENNE Cedex 1 FRANCE
LIN Peixiong	Université Paris VI, 4, Place Jussieu 75005 PARIS FRANCE
LUSKIN Mitchell	Institute for Math. and its Applications, University of Minnesota, Vincent Hall 514, Church Street S.E. MINNEAPOLIS, MN 55455 U.S.A
MODICA Luciano	Dipartimento di Matématica, Università di Pisa, Via Buonarroti, 2 56100 PISA ITALIE
MULLER Ingo	Technische Universität Berlin, Hermann-Föttinger Institut für Thermo und Fluiddynamik, Strasse des 17. juini 135, D 1000 Berlin 12 FEDERAL REPUBLIC OF GERMANY
MULLER Stephan	Laboratoire d'Analyse Numérique, Tour 55 65,Université Pierre et Marie Curie 75230 PARIS Cedex 05 FRANCE
NAVARD Patrick	Ecole des Mines Sophia-Antipolis 06565 VALBONNE Cedex FRANCE
NICOLAS Gérald	E.D.F. Direction des Etudes et Recherche, Service Réacteurs nucléaires et échangeurs Quai Watier - B.P. 49. 78401 CHATOU Cedex FRANCE
PEGO Robert	Department of Mathematics University of Michigan ANN ARBOR, MI 48 H 109 U.S.A.
PELCE Pierre	Faculté des Sciences et Techniques de Saint-Jérome Rue Henri Poincaré 13397 MARSEILLE Cedex 13 FRANCE
PENEL Patrick	Département de Mathématiques, Université de Toulon et du Var 83130 LA GARDE Cedex FRANCE
PLOHR Bradley	Computer Sciences Departement, University of Wisconsin MADISON, WI 53706 U.S.A.
PUEL Jean-Pierre	U.E.R. Sc. Fondamentales et Appliquées, Université d'Orléans Rue de Chartres; 45046 ORLEANS Cedex FRANCE
ROUX Bernard	Institut de Mécanique des Fluides, 1, Rue Honorat 13003 MARSEILLE FRANCE
SAUT Jean-Claude	Analyse Numérique, Bt 425, Université Paris-Sud, 91405 ORSAY FRANCE
SEKERKA Robert	Office of the Dean Mellon College of Science Carnegie-Mellon University PITTSBURGH, PA 15213-3890 U.S.A.
SERRE Denis	Laboratoire de Mathématiques, Ecole Normale Supérieure, 46, Allée d'Italie 69364 LYON Cedex 7 FRANCE
SHEARER Michael	Departement of Mathematics, Box 8205, North Carolina State University RALEIGH, N.C. 27695-8205 U.S.A.

SLEMROD Marshall	Departement of Mathematics, University of Wisconsin MADISON, WI 53706 U.S.A.
THOMPSON Philip	Department of Mechanical Engineering Aeronautical Engineering and Mechanics Rensselaer Polytechnic Institute TROY, N.Y. 12180-3590 U.S.A.
ZAVADA John	European Research Office of the U.S. Army. U.S.A.R.D.S.G. U.K. Old Marylebone Road. LONDON NW 1 5TH ENGLAND

Lecture Notes in Mathematics

Vol. 1236: Stochastic Partial Differential Equations and Applications. Proceedings, 1985. Edited by G. Da Prato and L. Tubaro. V, 257 pages. 1987.

Vol. 1237: Rational Approximation and its Applications in Mathematics and Physics. Proceedings, 1985. Edited by J. Gilewicz, M. Pindor and W. Siemaszko. XII, 350 pages. 1987.

Vol. 1250: Stochastic Processes – Mathematics and Physics II. Proceedings 1985. Edited by S. Albeverio, Ph. Blanchard and L. Streit. VI, 359 pages. 1987.

Vol. 1251: Differential Geometric Methods in Mathematical Physics. Proceedings, 1985. Edited by P.L. García and A. Pérez-Rendón. VII, 300 pages. 1987.

Vol. 1255: Differential Geometry and Differential Equations. Proceedings, 1985. Edited by C. Gu, M. Berger and R.L. Bryant. XII, 243 pages. 1987.

Vol. 1256: Pseudo-Differential Operators. Proceedings, 1986. Edited by H.O. Cordes, B. Gramsch and H. Widom. X, 479 pages. 1987.

Vol. 1258: J. Weidmann, Spectral Theory of Ordinary Differential Operators. VI, 303 pages. 1987.

Vol. 1260: N.H. Pavel, Nonlinear Evolution Operators and Semigroups. VI, 285 pages. 1987.

Vol. 1263: V.L. Hansen (Ed.), Differential Geometry. Proceedings, 1985. XI, 288 pages. 1987.

Vol. 1265: W. Van Assche, Asymptotics for Orthogonal Polynomials. VI, 201 pages. 1987.

Vol. 1267: J. Lindenstrauss, V.D. Milman (Eds.), Geometrical Aspects of Functional Analysis. Seminar. VII, 212 pages. 1987.

Vol. 1269: M. Shiota, Nash Manifolds. VI, 223 pages. 1987.

Vol. 1270: C. Carasso, P.-A. Raviart, D. Serre (Eds.), Nonlinear Hyperbolic Problems. Proceedings, 1986. XV, 341 pages. 1987.

Vol. 1272: M.S. Livšic, L.L. Waksman, Commuting Nonselfadjoint Operators in Hilbert Space. III, 115 pages. 1987.

Vol. 1273: G.-M. Greuel, G. Trautmann (Eds.), Singularities, Representation of Algebras, and Vector Bundles. Proceedings, 1985. XIV, 383 pages. 1987.

Vol. 1275: C.A. Berenstein (Ed.), Complex Analysis I. Proceedings, 1985–86. XV, 331 pages. 1987.

Vol. 1276: C.A. Berenstein (Ed.), Complex Analysis II. Proceedings, 1985–86. IX, 320 pages. 1987.

Vol. 1277: C.A. Berenstein (Ed.), Complex Analysis III. Proceedings, 1985–86. X, 350 pages. 1987.

Vol. 1283: S. Mardešić, J. Segal (Eds.), Geometric Topology and Shape Theory. Proceedings, 1986. V, 261 pages. 1987.

Vol. 1285: I.W. Knowles, Y. Saitō (Eds.), Differential Equations and Mathematical Physics. Proceedings, 1986. XVI, 499 pages. 1987.

Vol. 1287: E.B. Saff (Ed.), Approximation Theory, Tampa. Proceedings, 1985–1986. V, 228 pages. 1987.

Vol. 1288: Yu. L. Rodin, Generalized Analytic Functions on Riemann Surfaces. V, 128 pages, 1987.

Vol. 1294: M. Queffélec, Substitution Dynamical Systems – Spectral Analysis. XIII, 240 pages. 1987.

Vol. 1299: S. Watanabe, Yu.V. Prokhorov (Eds.), Probability Theory and Mathematical Statistics. Proceedings, 1986. VIII, 589 pages. 1988.

Vol. 1300: G.B. Seligman, Constructions of Lie Algebras and their Modules. VI, 190 pages. 1988.

Vol. 1302: M. Cwikel, J. Peetre, Y. Sagher, H. Wallin (Eds.), Function Spaces and Applications. Proceedings, 1986. VI, 445 pages. 1988.

Vol. 1303: L. Accardi, W. von Waldenfels (Eds.), Quantum Probability and Applications III. Proceedings, 1987. VI, 373 pages. 1988.

Lecture Notes in Physics

Vol. 321: J. Pitowsky, Quantum Probability – Quantum Logic. IX, 209 pages. 1989.

Vol. 322: M. Schlichenmaier, An Introduction to Riemann Surfaces, Algebraic Curves and Moduli Spaces. XIII, 148 pages. 1989.

Vol. 323: D.L. Dwoyer, M.Y. Hussaini, R.G. Voigt (Eds.), 11th International Conference on Numerical Methods in Fluid Dynamics. XIII, 622 pages. 1989.

Vol. 324: P. Exner, P. Šeba (Eds.), Applications of Self-Adjoint Extensions in Quantum Physics. Proceedings, 1987. VIII, 273 pages. 1989.

Vol. 325: E. Brändas, N. Elander (Eds.), Resonances, Proceedings, 1987. XVIII, 564 pages. 1989.

Vol. 326: A. Grauel, Feldtheoretische Beschreibung der Thermodynamik für Grenzflächen. IX, 317 Seiten. 1989.

Vol. 327: K. Meisenheimer, H.-J. Röser (Eds.), Hot Spots in Extragalactic Radio Source. Proceedings, 1988, XII, 301 pages. 1989.

Vol. 328: G. Wegner (Ed.), White Dwarfs. Proceedings, 1988. XIV, 524 pages. 1989.

Vol. 329: A Heck, F. Murtagh (Eds.), Knowledge Based Systems in Astronomy. IV, 280 pages. 1989.

Vol. 330: J.M. Moran, J.N. Hewitt, K.Y. Lo (Eds.), Gravitational Lenses. Proceedings, 1988. XIV, 238 pages. 1989.

Vol. 331: G. Winnewisser, J.T. Armstrong (Eds.), The Physics and Chemistry of Interstellar Molecular Clouds mm and Sub-mm Observations in Astrophysics. Proceedings, 1988. XVIII, 463 pages. 1989.

Vol. 332: P. Flin, H.W. Duerbeck (Eds.), Morphological Cosmology. Proceedings, 1988. VII, 438 pages. 1989.

Vol. 333: I. Appenzeller, H.J. Habing, P. Léna (Eds.), Evolution of Galaxies – Astronomical Observations. Proceedings, 1988. X, 391 pages. 1989.

Vol. 334: L. Maraschi, T. Maccacaro, M.-H. Ulrich (Eds.), BL Lac Objects. XIII, 497 pages. 1989.

Vol. 335: A. Lakhtakia, V.K. Varadan, V.V. Varadan, Time-Harmonic Electromagnetic Fields in Chiral Media. VII, 121 pages. 1989.

Vol. 336: M. Müller, Consistent Classical Supergravity Theories. VI, 125 pages. 1989.

Vol. 337: A.P. Maclin, T.L. Gill, W.W. Zachary (Eds.), Magnetic Phenomena. Proceedings, 1988. VI, 142 pages. 1989.

Vol. 338: V. Privman, N.M. Švrakić, Directed Models of Polymers, Interfaces, and Clusters: Scaling and Finite-Size Properties. VI, 120 pages. 1989.

Vol. 339: F. Ehlotzky (Ed.), Fundamentals of Laser Interactions II. Proceedings, 1989. XI, 317 pages. 1989.

Vol. 340: M. Peshkin, A. Tonomura, The Aharonov-Bohm Effect. VI, 152 pages. 1989.

Vol. 341: E.F. Milone (Ed.), Infrared Extinction and Standardization. Proceedings, 1988. III, 79 pages. 1989.

Vol. 342: M. Balabane, P. Lochak, C. Sulem (Eds.), Integrable Systems and Applications. Proceedings, 1988. VII, 342 pages. 1989.

Vol. 343: M. Month, S. Turner (Eds.), Frontiers of Particle Beams; Observation, Diagnosis and Correction. Proceedings, 1988. IX, 509 pages. 1989.

Vol. 344: M. Rascle, D. Serre, M. Slemrod (Eds.), PDEs and Continuum Models of Phase Transitions. Proceedings, 1988. VII, 229 pages. 1989.